Core Books in Advanced Mathematics

Integration

D1427768

Core Books in Advanced Mathematics

General Editor: C. PLUMPTON, Moderator in Mathematics,
University of London School Examinations Department;
formerly Reader in Engineering Mathematics,
Queen Mary College, University of London.

Advisory Editor: N. WARWICK

Titles available:

Differentiation
Integration
Vectors
Curve Sketching

Core Books in Advanced Mathematics

Integration

C. T. Moss
Formerly Chief Examiner and Moderator in Advanced Level
Mathematics, AEB; Senior Lecturer and Deputy Head of
Mathematics, The City University, London.

C. Plumpton
Moderator in Mathematics, University of
London School Examinations Department;
formerly Reader in Engineering Mathematics,
Queen Mary College, University of London.

Macmillan Education
London and Basingstoke

First published 1983

Published by
Macmillan Education Limited
Houndmills Basingstoke Hampshire RG21 2XS
and London
Associated companies throughout the world

Printed in Hong Kong

British Library Cataloguing in Publication Data
Moss, C. T.
Integration. – (Core books in advanced mathematics)
1. Integrals
I. Title   II. Plumpton, Charles
III. Series
515.4'3   QA308
ISBN 0-333-31793-9

Contents

Preface

Advanced level mathematics syllabuses are once again undergoing changes of content and approach, following the revolution in the early 1960s which led to the unfortunate dichotomy between 'modern' and 'traditional' mathematics. The current trend in syllabuses for Advanced level mathematics now being developed and published by many GCE Boards is towards an integrated approach, taking the best of the topics and approaches of the modern and traditional, in an attempt to create a realistic examination target, through syllabuses which are maximal for examining and minimal for teaching. In addition, resulting from a number of initiatives, core syllabuses are being developed for Advanced level mathematics syllabuses, consisting of techniques of pure mathematics as taught in schools and colleges at this level.

The concept of a core can be used in several ways, one of which is mentioned above, namely the idea of a core syllabus to which options such as theoretical mechanics, further pure mathematics and statistics can be added. The books in this series are core books involving a different use of the core idea. They are books on a range of topics, each of which is central to the study of Advanced level mathematics; they form small core studies of their own, of topics which together cover the main areas of any single-subject mathematics syllabus at Advanced level.

Particularly at times when economic conditions make the problems of acquiring comprehensive textbooks giving complete syllabus coverage acute, schools and colleges and individual students can collect as many of the core books as they need, one or more, to supplement books already possessed, so that the most recent syllabuses of, for example, the London, Cambridge, AEB and JMB GCE Boards, can be covered at minimum expense. Alternatively, of course, the whole set of core books give complete syllabus coverage of single-subject Advanced level mathematics syllabuses.

The aim of each book is to develop a major topic of the single-subject syllabuses, giving essential book work and worked examples and exercises arising from the authors' vast experience of examining at this level and including actual past GCE questions also. Thus, as well as using the core books in either of the above ways, they would also be ideal for supplementing comprehensive textbooks in the sense of providing more examples and exercises, so necessary for preparation and revision for examinations on the Advanced level mathematics syllabuses offered by the GCE Boards.

This book on integration follows closely the ideas developed in the companion book on differentiation, and is intended to be read in conjunction with that book.

Integration is treated as the reverse process to differentiation and, as such, the basic results are tabulated. From them a more generalised table of results is derived and the first six chapters are devoted to ensuring that the reader becomes familiar with the accepted techniques of integration – i.e. integration by substitution, the involvement of trigonometric functions, rational and irrational algebraic fractions and integration by parts. Definite integrals and the properties of definite integrals are also discussed and, provided the readers follow carefully the many worked examples and try for themselves the complementary exercises, they should experience little difficulty in coping with the integrals to be found in most GCE single-mathematics examination papers. The authors also believe that many first-year university and polytechnic engineering and applied science students will find the book a valuable aid to ensuring proficiency in integration.

Applications of integration, approximate methods of integration and simple first-order variable separable type differential equations are also discussed in the second part of the book.

The authors are grateful to the following GCE Examining Boards for permission to reproduce questions from past Advanced Level GCE papers: University of London Entrance and School Examinations Council (L); The Associated Examining Board (AEB).

C. T. Moss
C. Plumpton

1 Basic definitions and results

1.1 Definitions

In *Differentiation* we learned how to differentiate a function of the variable x. Frequently we require to carry out the reverse procedure, i.e. given the derivative, or the rate of change, of a function f, we need to find the function f. This reverse process is known as integration and, in the first instance, we define it as the inverse of differentiation.

For a function $F(x)$ defined by $\dfrac{dF(x)}{dx} = f(x)$, we call $F(x)$ the integral of the integrand $f(x)$ and write

$$F(x) = \int f(x)\,dx.$$

Strictly speaking, since the derivative of a constant is zero, we should say $\int f(x)\,dx = F(x) + C$, where C is an arbitrary constant known as the *constant of integration*. For convenience we shall drop the arbitrary constant, introducing it only when it is essential to the problem in question. Also, we assume that the values of x are such as to make the result meaningful, for example writing $\ln x$ implies $x > 0$.

Hence, using our table of known standard derivatives, we can draw up a table of standard integrals.

$F(x)$	$\dfrac{dF(x)}{dx} = f(x)$	$\Rightarrow$	$g(x)$	$\int g(x)\,dx$
x^n	nx^{n-1}	$\Rightarrow$	$x^n, n \neq -1$	$\dfrac{x^{n+1}}{n+1}$
$\ln x$	$\dfrac{1}{x}$	$\Rightarrow$	$\dfrac{1}{x}$	$\ln x$
e^x	e^x	$\Rightarrow$	e^x	e^x
$\sin x$	$\cos x$	$\Rightarrow$	$\cos x$	$\sin x$
$\cos x$	$-\sin x$	$\Rightarrow$	$\sin x$	$-\cos x$
$\tan x$	$\sec^2 x$	$\Rightarrow$	$\sec^2 x$	$\tan x$

If in doubt about any of these results, you should try differentiating back using the results of the first two columns, e.g.

$$\frac{d}{dx}(-\cos x) = -\frac{d}{dx}(\cos x) = -(-\sin x) = \sin x.$$

It is essential that you should commit these six standard integrals to memory; from them we shall deduce many more complicated integrals.

1.2   Sum or difference of functions

Since the derivative of the sum (or difference) of a finite number of functions is the sum (or difference) of their individual derivatives, it follows conversely that the integral of the sum (or difference) of a finite number of functions is the sum (or difference) of their integrals taken separately.

Example 1

(i) $\displaystyle\int (x^3 + \sin x)\,dx = \int x^3\,dx + \int \sin x\,dx = \frac{x^4}{4} - \cos x,$

(ii) $\displaystyle\int \left(\frac{1}{\sqrt[3]{x}} - \frac{1}{x}\right)dx = \int x^{-1/3}\,dx - \int \frac{1}{x}\,dx = \frac{x^{2/3}}{2/3} - \ln x = \frac{3x^{2/3}}{2} - \ln x,$

(iii) $\displaystyle\int (1 - 2x)^3\,dx = \int (1 - 6x + 12x^2 - 8x^3)\,dx = x - 3x^2 + 4x^3 - 2x^4.$

Remember that, strictly speaking, an arbitrary constant of integration should be added in each case.

1.3   Two (elementary) rules

THEOREM I   If $\dfrac{d}{dx}F(x) = f(x)$, then $\displaystyle\int f(A + x)\,dx = F(A + x)$, where A is any constant; i.e. the addition of a constant to the variable wherever it occurs makes no difference to the form of the integral.

Example 2

(i) $\displaystyle\int \sin x\,dx = -\cos x \Rightarrow \int \sin (x + \pi/2)\,dx = -\cos (x + \pi/2),$

(ii) $\displaystyle\int x^2\,dx = \frac{x^3}{3} \Rightarrow \int (a + x)^2\,dx = \frac{(a + x)^3}{3},$

(iii) $\displaystyle\int \frac{1}{x}\,dx = \ln x \Rightarrow \int \frac{1}{c + x}\,dx = \ln (c + x).$

Care must be taken when using this rule. It only applies when x is replaced by $A + x$ and does not cover the cases when the expression in the brackets is not linear, i.e. not of the first degree in x. For example

$$\int (a + x^2)^2 \, dx \text{—this is NOT } \frac{(a + x^2)^3}{3}.$$

(This integral should be determined by first multiplying out the bracket and then integrating.)

The proof of Theorem I follows directly from the derivative of a function of a function and the fact that $\frac{d}{dx}(A + x) = 1$.

Thus for $\frac{d}{dx} F(A + x)$, let $A + x = t$

$$\Rightarrow \frac{dt}{dx} = 1 \quad \text{and} \quad \frac{d}{dx} F(A + x) = \frac{dt}{dx} \frac{d}{dt} F(t)$$

$$= 1 . \frac{d}{dt} F(t) = f(t), \text{ say, } = f(A + x)$$

$$\Rightarrow \int f(A + x) \, dx = F(A + x).$$

Example 3

(i) $\displaystyle \int \sqrt{(3 + x)} \, dx = \frac{(3 + x)^{3/2}}{3/2} = \frac{2}{3}(3 + x)^{3/2},$

(ii) $\displaystyle \int \sec^2 (\pi/3 + x) \, dx = \tan (\pi/3 + x),$

(iii) $\displaystyle \int \left[\cos (x + \pi/6) + \frac{1}{2 + x} + (5 + x)^{5/2} \right] dx$

$$= \sin (x + \pi/6) + \ln (2 + x) + \frac{2}{7}(5 + x)^{7/2}.$$

THEOREM II If $\dfrac{d}{dx} F(x) = f(x)$, then $\displaystyle \int f(Bx + A) \, dx = \frac{1}{B} F(Bx + A)$, where A and B are constants, i.e., if x is replaced by $Bx + A$, then the form of the integral remains the same but the answer must be divided by the constant B.

Example 4

(i) $\displaystyle \int x^3 \, dx = \frac{x^4}{4} \Rightarrow \int (2x + 5)^3 \, dx = \left(\frac{1}{2}\right)\left(\frac{1}{4}\right)(2x + 5)^4 = \frac{1}{8}(2x + 5)^4,$

(ii) $\displaystyle \int \sin 3x \, dx = \frac{-\cos 3x}{3},$

(iii) $\int \dfrac{1}{3-4x}dx = \int \dfrac{1}{3+(-4)x}dx = \dfrac{1}{-4}\ln(3-4x),$

(iv) $\int e^{ax}\,dx = \dfrac{1}{a}e^{ax},\ a$ being a constant.

A more generalised table of the six standard integrals can now be formed.

$f(x),\ a \neq 0$	$\int f(x)\,dx$
$(ax+b)^n,\ n \neq -1$	$\dfrac{(ax+b)^{n+1}}{a(n+1)}$
$\dfrac{1}{ax+b}$	$\dfrac{1}{a}\ln(ax+b)$
e^{ax+b}	$\dfrac{1}{a}e^{ax+b}$
$\sin ax$	$-\dfrac{1}{a}\cos ax$
$\cos ax$	$\dfrac{1}{a}\sin ax$
$\sec^2 ax$	$\dfrac{1}{a}\tan ax$

The proof of Theorem II again follows directly from the derivative of a function of a function.

$$\frac{d}{dx}F(Bx+A) = \frac{dt}{dx}\cdot\frac{d}{dt}F(t), \quad \text{where } t = Bx+A, = Bf(t),\ \text{say.}$$

Rearranging,

$$f(t) = f(Bx+A) = \frac{1}{B}\frac{d}{dx}F(Bx+A)$$

$$\Rightarrow \int f(Bx+A)\,dx = \frac{1}{B}F(Bx+A).$$

Exercise 1.3

Integrate with respect to the appropriate variable. Check your answer by differentiating the expression obtained.

1 $(7+x)^2,$ **2** $\sqrt{(3-2x)},$ **3** $\sqrt[3]{(1-t)},$ **4** $\left(1+\dfrac{x}{2}\right)^7,$ **5** $(3x-2)^{13/6},$ **6** $\dfrac{1}{\sqrt[4]{(3-2u)}},$

7 $\dfrac{1}{1-x},$ **8** $\dfrac{1}{2x+3},$ **9** $\sin 2x,$ **10** $\sin(\pi/2 - t),$ **11** $\sin[3(\pi+x)],$ **12** $\cos 4x,$

13 $\cos \frac{1}{2}t$, **14** $\sec^2 4x$, **15** $\sec^2 3\left(\pi - \frac{u}{4}\right)$, **16** e^{2x}, **17** e^{1-2x}, **18** $\frac{1}{e^{3u}}$,

19 $(e^x + e^{-x})^2$, **20** $(2 + e^{3x})^2$.

1.4 Extension of the list of standard integrals; Inverse trigonometric functions

Using the results of further known derivatives we can extend the table of standard integrals to include the inverse trigonometric functions.

$$\frac{d}{dx} \sin^{-1} x = \frac{1}{\sqrt{(1 - x^2)}} \Rightarrow \int \frac{1}{\sqrt{(1 - x^2)}} dx = \sin^{-1} x,$$

or more generally

$$\frac{d}{dx} \sin^{-1}\left(\frac{x}{a}\right) = \frac{1}{\sqrt{(a^2 - x^2)}} \Rightarrow \int \frac{1}{\sqrt{(a^2 - x^2)}} dx = \sin^{-1}\left(\frac{x}{a}\right), \quad a > 0.$$

Thus a more extended table includes

$f(x)$	$\int f(x)\,dx$
$\dfrac{1}{\sqrt{(a^2 - x^2)}}, a > 0$	$\sin^{-1}\left(\dfrac{x}{a}\right)$
$-\dfrac{1}{\sqrt{(a^2 - x^2)}}, a > 0$	$\cos^{-1}\left(\dfrac{x}{a}\right)$
$\dfrac{1}{a^2 + x^2}, \quad a > 0$	$\dfrac{1}{a}\tan^{-1}\left(\dfrac{x}{a}\right)$

Example 5

(i) $\displaystyle\int \frac{1}{\sqrt{(9 - x^2)}} dx = \sin^{-1}\left(\frac{x}{3}\right),$

(ii) $\displaystyle\int \frac{1}{\sqrt{(9 - 4x^2)}} dx = \frac{1}{2}\int \frac{1}{\sqrt{[(3/2)^2 - x^2]}} dx,$

which is now expressed in standard form and gives

$$\int \frac{1}{\sqrt{(9 - 4x^2)}} dx = \frac{1}{2}\sin^{-1}\left[\frac{x}{(3/2)}\right] = \frac{1}{2}\sin^{-1}(2x/3),$$

(iii) $\displaystyle\int \frac{1}{1 + 4x^2} dx = \frac{1}{4}\int \frac{1}{(\frac{1}{2})^2 + x^2} dx = \frac{1}{4}\cdot\frac{1}{(\frac{1}{2})}\tan^{-1}\left[\frac{x}{(\frac{1}{2})}\right] = \frac{1}{2}\tan^{-1}(2x).$

This last result could have been deduced immediately by the use of Theorem II and the standard form $\displaystyle\int \frac{1}{a^2 + x^2} dx = \frac{1}{a}\tan^{-1}\left(\frac{x}{a}\right)$. Thus $\displaystyle\int \frac{1}{1 + (2x)^2} dx$ is of the standard form where $a = 1$ and the variable x has been replaced by $2x$. Therefore the form of the integral remains the same but the answer must be divided by 2, i.e.

$$\int \frac{1}{1 + (2x)^2} dx = \frac{\frac{1}{2}\tan^{-1}(2x)}{2} = \frac{1}{2}\tan^{-1}(2x).$$

(iv) $\int \frac{1}{29 + 4x + x^2} dx = \int \frac{1}{25 + (x+2)^2} dx = \int \frac{1}{5^2 + (x+2)^2} dx$

which is now of the form $\int \frac{1}{a^2 + x^2} dx = \frac{1}{a}\tan^{-1}\left(\frac{x}{a}\right)$, where $a = 5$ and x has been replaced by $x + 2$. Hence, by Theorem I,

$$\int \frac{1}{29 + 4x + x^2} dx = \frac{1}{5}\tan^{-1}\left(\frac{x+2}{5}\right).$$

We shall reconsider this form of integral later, but it demonstrates that in integration we have to rely heavily on our experience of recognising a derivative and/or a standard result in disguised form.

Exercise 1.4

Integrate: **1** $\dfrac{1}{\sqrt{(4 - x^2)}}$, **2** $\dfrac{1}{\sqrt{[4 - (x+1)^2]}}$, **3** $\dfrac{1}{\sqrt{(25 - 9x^2)}}$, **4** $\dfrac{1}{100 + x^2}$,

5 $\dfrac{1}{1 + 100x^2}$, **6** $\dfrac{1}{3x^2 + 5}$, **7** $\dfrac{1}{\sqrt{[4 - (x-3)^2]}}$, **8** $\dfrac{1}{17 + 4x + 4x^2}$.

2   Integration by substitution

2.1   The basic method

Some integrals fall into clearly defined recognisable categories for which we have established definite rules of procedure. Many others have to be converted into recognisable forms either by experience or by changing the variable. Integration by the method of substitution is essentially the converse of differentiation of a function of a function and, as such, depends upon the use of the chain rule.

Consider $y = \int f(x)\,dx$. Then $\dfrac{dy}{dx} = f(x)$ and, if a substitution is made so that x is a function of t, we can write the integral in the form $y = \int f(x)\dfrac{dx}{dt}\,dt$.

This, when expressed entirely in terms of t, may reduce to a known standard form which can be integrated on sight.

Example 1

(i) $\displaystyle\int (1 + x^2)^{3/2} x\,dx$. Let $1 + x^2 = t$,

$$\frac{dt}{dx} = 2x \Rightarrow \frac{dx}{dt} = \frac{1}{2x}$$

$$\Rightarrow \int (1 + x^2)^{3/2} x\,dx = \int t^{3/2} . x . \frac{1}{2x}\,dt = \frac{1}{2}\int t^{3/2}\,dt$$

$$= \frac{1}{2}.\frac{2}{5}t^{5/2} = \frac{1}{5}(1 + x^2)^{5/2},$$

(ii) $\displaystyle\int \sin^3 x \cos x\,dx$. Let $\sin x = t$,

$$\frac{dt}{dx} = \cos x \Rightarrow \frac{dx}{dt} = \frac{1}{\cos x}$$

$$\Rightarrow \int \sin^3 x \cos x\,dx = \int t^3 \cos x \frac{1}{\cos x}\,dt = \int t^3\,dt$$

$$= \frac{t^4}{4} = \frac{\sin^4 x}{4},$$

(iii) $\displaystyle\int \frac{x^2}{1 + x^6}\,dx.$ Let $x^3 = t,$

$$\frac{dt}{dx} = 3x^2 \Rightarrow \frac{dx}{dt} = \frac{1}{3x^2}$$

$$\Rightarrow \int \frac{x^2}{1 + x^6}\,dx = \int \frac{x^2}{1 + t^2} \cdot \frac{1}{3x^2}\,dt$$

$$= \frac{1}{3}\int \frac{1}{1 + t^2}\,dt$$

$$= \frac{1}{3}\tan^{-1} t = \frac{1}{3}\tan^{-1}(x^3).$$

It is clear from all three examples that, in choosing a substitution, one must look carefully at the integrand and try to separate from it a part which is related by differentiation to all or part of the remaining piece of the integrand. Thus,

$$\text{for } \int (1 + x^2)^{3/2} x\,dx, \quad \text{we see } \frac{d}{dx}(1 + x^2) = 2x,$$

$$\text{for } \int \sin^3 x \cos x\,dx, \quad \text{we see } \frac{d}{dx}(\sin x) = \cos x,$$

$$\text{for } \int \frac{x^2}{1 + x^6}\,dx, \quad \text{we see } \frac{d}{dx}(x^3) = 3x^2.$$

Only experience gained by solving many such integrals by this method will lead to an automatic choice of the correct substitution and the student is therefore advised to try as many as possible. In each of the following examples a lead is given. Use it to integrate the given function. Then do Exercise 2.1.

Example 2

(i) $\displaystyle\frac{\sin 2x}{1 - \cos 2x}$ $\displaystyle\frac{d}{dx}(\cos 2x) = -2\sin 2x \Rightarrow t = \cos 2x,$

$$\text{(or } t = 1 - \cos 2x),$$

(ii) $\displaystyle\frac{e^x}{1 + e^{2x}}$ $\displaystyle\frac{d}{dx}(e^x) = e^x \Rightarrow t = e^x,$

(iii) $\displaystyle\frac{1}{x}(1 + \ln x)^3$ $\displaystyle\frac{d}{dx}(\ln x) = \frac{1}{x} \Rightarrow t = \ln x,$

(iv) $\sec^2 3x \tan^3 3x$ $\displaystyle\frac{d}{dx}(\tan 3x) = 3\sec^2 3x \Rightarrow t = \tan 3x,$

(v) $\displaystyle\frac{\cos 2x}{\sqrt{(1 - \sin^2 2x)}}$ $\displaystyle\frac{d}{dx}(\sin 2x) = 2\cos 2x \Rightarrow t = \sin 2x.$

Exercise 2.1

Integrate: **1** $\dfrac{\cos 2x}{1 + \sin 2x}$, **2** $\sin^5 x \cos x$, **3** $\dfrac{\sin^{-1} x}{\sqrt{(1 - x^2)}}$, **4** $\dfrac{x}{1 + 2x^2}$, **5** $\dfrac{x}{\sqrt{(9 - 4x^2)}}$,

6 $\dfrac{(\ln x)^2}{x}$, **7** $\dfrac{1}{x \ln x}$, **8** $\dfrac{e^{3x}}{e^{3x} - 1}$, **9** $\dfrac{\sec^2 x}{(1 - \tan x)^2}$, **10** $\dfrac{\sin \sqrt{x}}{\sqrt{x}}$, **11** $\dfrac{x}{\sqrt{(x - 4)}}$,

12 $\dfrac{\sin 4x}{\cos^3 4x}$, **13** $\dfrac{x^3}{1 + x^8}$, **14** $x\sqrt{(1 + 4x^2)}$, **15** $\dfrac{\cos x}{\sqrt{(\sin x)}}$, **16** $\dfrac{e^x}{(1 + e^x)^3}$,

17 $\sin x \, e^{\cos x}$, **18** $\sin 2x \, e^{\cos^2 x}$, **19** $\dfrac{\cos x}{\sqrt{(\sin x + 1)}}$.

2.2 Further substitutions

Unfortunately, it is not always easy to pick out the derivative part of an integrand and in such cases one can only resort to experience gained from other sections of mathematics such as trigonometry. Even then one sometimes has to make a guess substitution and hope that it will ultimately lead to a standard form. However, with integrals involving square roots, one should always try to make a substitution which will enable the square root sign to be eliminated. Here the trigonometric formulae can often prove useful.

Example 3

(i) $\displaystyle\int \sqrt{(4 - x^2)}\,dx$.

The square root sign can be eliminated by making the expression under the sign a perfect square. This can be done using the substitution $x = 2 \sin \theta$, for then

$$\sqrt{(4 - x^2)} = \sqrt{(4 - 4\sin^2 \theta)} = \sqrt{[4(1 - \sin^2 \theta)]} = \sqrt{(4 \cos^2 \theta)} = 2 \cos \theta.$$

Also
$$x = 2 \sin \theta \Rightarrow \frac{dx}{d\theta} = 2 \cos \theta$$

$$\Rightarrow \int \sqrt{(4 - x^2)}\,dx = \int 2 \cos \theta \cdot 2 \cos \theta \, d\theta = \int 4 \cos^2 \theta \, d\theta,$$

(which can now be determined using the trigonometric formula $\cos 2A = 2 \cos^2 A - 1$, i.e. $2 \cos^2 A = 1 + \cos 2A$),

$$\Rightarrow \int \sqrt{(4 - x^2)}\,dx = \int 2(1 + \cos 2\theta)\,d\theta = 2\left[\theta + \frac{1}{2}\sin 2\theta\right]$$

$$= 2[\theta + \sin \theta \cos \theta]$$

$$= 2\left[\sin^{-1}\left(\frac{x}{2}\right) + \frac{x}{2}\sqrt{\left(1 - \frac{x^2}{4}\right)}\right]$$

$$= 2 \sin^{-1}\left(\frac{x}{2}\right) + \frac{1}{2}x\sqrt{(4 - x^2)},$$

(ii) $\int \dfrac{1}{x + \sqrt{x}}\, dx.$

Here again we could eliminate the square root sign by making a trigonometric substitution, $x = \sin^2 \theta$, but this leads to an integral which we do not easily recognise.

However, in this case the substitution $x = t^2$ would be just as good, if not better.

$$x = t^2 \Rightarrow \dfrac{dx}{dt} = 2t$$

$$\Rightarrow \int \dfrac{1}{x + \sqrt{x}}\, dx = \int \dfrac{1}{t^2 + \sqrt{t^2}} \cdot 2t\, dt = \int \dfrac{2}{t + 1}\, dt$$

$$= 2 \ln (t + 1) = 2 \ln (\sqrt{x} + 1),$$

(iii) $\int \dfrac{1}{x^2 \sqrt{(1 + x^2)}}\, dx.$

$$x = \tan \theta \Rightarrow \dfrac{dx}{d\theta} = \sec^2 \theta$$

$$\Rightarrow \int \dfrac{1}{x^2 \sqrt{(1 + x^2)}}\, dx = \int \dfrac{\sec^2 \theta}{\tan^2 \theta \sqrt{(1 + \tan^2 \theta)}}\, d\theta$$

$$= \int \dfrac{\sec^2 \theta}{\tan^2 \theta . \sec \theta}\, d\theta, \quad \text{since } 1 + \tan^2 \theta = \sec^2 \theta,$$

$$= \int \dfrac{\cos \theta}{\sin^2 \theta}\, d\theta = \int \dfrac{d}{d\theta}\left(-\dfrac{1}{\sin \theta}\right) d\theta = -\dfrac{1}{\sin \theta}.$$

But $x = \tan \theta \Rightarrow \sin \theta = \dfrac{x}{\sqrt{(1 + x^2)}}.$

Hence $$\int \dfrac{1}{x^2 \sqrt{(1 + x^2)}}\, dx = -\dfrac{\sqrt{(1 + x^2)}}{x}.$$

Exercise 2.2

Integrate: 1 $\dfrac{x}{\sqrt{(1 - x^2)}},$ 2 $\dfrac{1}{(4 + x^2)^{3/2}},$ 3 $x\sqrt{(x - 1)},$ 4 $\dfrac{1}{(1 - x)^{3/2}}.$

2.3 Alternative notation

When it is easy to see the derivative part of the integrand and consequently the required substitution, it is sometimes possible to perform the integration without actually making the substitution.

Consider $$\int \sin^n x \cos x\, dx, \quad n \neq -1.$$

Clearly the derivative of sin x is cos x. This can be written as $\dfrac{d}{dx}(\sin x) = \cos x$, or, when used in the integral, $d(\sin x) \equiv \cos x \, dx$.

Thus we can write $\displaystyle\int \sin^n x \cos x \, dx = \int \sin^n x \, d(\sin x)$ and read it as 'integrate $\sin^n x$ with respect to sin x', the same form of integral as $\displaystyle\int x^n \, dx$.

Thus $\displaystyle\int \sin^n x \cos x \, dx = \int \sin^n x \, d(\sin x) = \dfrac{\sin^{n+1} x}{n+1}$ for $n \neq -1$.

Generally such an integral can be written as

$$\int f^n(x) . f'(x) \, dx = \int f^n(x) \, d[f(x)] = \frac{f^{n+1}(x)}{n+1}.$$

Other standard forms in this notation that will be found useful include

$$\int \frac{f'(x)}{f(x)} \, dx = \int \frac{1}{f(x)} \, d[f(x)] = \ln f(x),$$

$$\int f'(x) e^{f(x)} \, dx = \int e^{f(x)} \, d[f(x)] = e^{f(x)},$$

$$\int \frac{f'(x)}{\sqrt{\{1 - [f(x)]^2\}}} \, dx = \int \frac{1}{\sqrt{\{1 - [f(x)]^2\}}} \, d[f(x)] = \sin^{-1} f(x).$$

Example 4

(i) $\displaystyle\int \frac{\sin 2x}{\cos^3 2x} \, dx = \int \frac{(-\tfrac{1}{2})(-2 \sin 2x)}{\cos^3 2x} \, dx$

$\displaystyle = \left(-\frac{1}{2}\right) \int \frac{1}{\cos^3 2x} \, d(\cos 2x)$

$\displaystyle = \left(-\frac{1}{2}\right) \frac{\cos^{-2} 2x}{-2} = \frac{1}{4 \cos^2 2x},$

(ii) $\displaystyle\int x^2 e^{2x^3} \, dx = \frac{1}{6} \int e^{2x^3} \, d(2x^3) = \frac{1}{6} e^{2x^3},$

(iii) $\displaystyle\int \frac{3}{1 - 4x} \, dx = -\frac{3}{4} \int \frac{1}{1 - 4x} \, d(1 - 4x) = -\frac{3}{4} \ln (1 - 4x),$

(iv) $\displaystyle\int \frac{x}{\sqrt{(1 - 2x^2)}} \, dx = -\frac{1}{4} \int \frac{1}{\sqrt{(1 - 2x^2)}} \, d(1 - 2x^2)$

$\displaystyle = -\frac{1}{4} \frac{\sqrt{(1 - 2x^2)}}{\frac{1}{2}} = -\frac{1}{2}\sqrt{(1 - 2x^2)},$

(v) $\displaystyle\int \frac{\cos \sqrt{x}}{\sqrt{x}} dx = 2 \int \cos \sqrt{x}\, d(\sqrt{x}) = 2 \sin \sqrt{x},$

(vi) $\displaystyle\int \frac{x}{\sqrt{(1 - x^4)}} dx = \frac{1}{2} \int \frac{1}{\sqrt{[1 - (x^2)^2]}} d(x^2) = \frac{1}{2} \sin^{-1} x^2.$

Exercise 2.3

Use the notation on p. 11 to integrate: **1** $\dfrac{\ln x}{x}$, **2** $\tan 3x \sec^2 3x$, **3** $\dfrac{x}{\sqrt[3]{(x^2 + 1)}}$,

4 $\dfrac{2x}{1 + 5x^2}$, **5** $\dfrac{e^x}{1 + e^x}$, **6** $e^{2x}(1 - e^{2x})^4$ **7** $\sin^3 x \cos x$, **8** $\tan 2x$, **9** $\dfrac{x^2}{\sqrt{(9 - x^6)}}$,

10 $\dfrac{\sin^{-1} x}{\sqrt{(1 - x^2)}}$, **11** $\dfrac{\sec^2 2x}{1 - \tan 2x}$, **12** $\dfrac{1}{x(1 + \ln x)^2}$.

3 Integrations involving trigonometric functions

3.1 Standard forms

So far the list of standard integrals involving trigonometric functions has been very limited, namely to

$$\int \sin ax \, dx = -\frac{1}{a} \cos ax,$$

$$\int \cos ax \, dx = \frac{1}{a} \sin ax,$$

$$\int \sec^2 ax \, dx = \frac{1}{a} \tan ax.$$

We can now extend this list to include

$$\int \tan ax \, dx = \int \frac{\sin ax}{\cos ax} dx = -\frac{1}{a} \int \frac{1}{\cos ax} d(\cos ax)$$

$$= -\frac{1}{a} \ln \cos ax = \frac{1}{a} \ln \sec ax,$$

$$\int \cot ax \, dx = \int \frac{\cos ax}{\sin ax} dx = \frac{1}{a} \int \frac{1}{\sin ax} d(\sin ax) = \frac{1}{a} \ln \sin ax,$$

$$\int \mathrm{cosec}\, ax \, dx = \int \frac{1}{\sin ax} dx = \int \frac{1}{2 \sin (ax/2) \cos (ax/2)} dx = \int \frac{\frac{1}{2} \sec^2 (ax/2)}{\tan (ax/2)} dx$$

$$= \frac{1}{a} \int \frac{1}{\tan (ax/2)} d[\tan (ax/2)] = \frac{1}{a} \ln \tan \frac{ax}{2},$$

$$\int \sec ax \, dx = \int \frac{1}{\cos ax} dx = \int \frac{\cos ax}{\cos^2 ax} dx = \frac{1}{a} \int \frac{1}{1 - \sin^2 ax} d(\sin ax)$$

$$= \frac{1}{2a} \int \left[\frac{1}{1 + \sin ax} + \frac{1}{1 - \sin ax} \right] d(\sin ax)$$

$$= \frac{1}{2a} [\ln (1 + \sin ax) - \ln (1 - \sin ax)]$$

$$= \frac{1}{2a} \ln \left(\frac{1 + \sin ax}{1 - \sin ax} \right).$$

This last result can be expressed in the alternative form
$$\int \sec ax\, dx = \frac{1}{a} \ln \tan \left[\frac{\pi}{4} + \frac{ax}{2} \right], \text{ or it could be deduced from}$$

$$\int \csc ax\, dx = \frac{1}{a} \ln \tan \frac{ax}{2} \text{ since } \int \sec ax\, dx = \int \csc (\pi/2 + ax)\, dx,$$

and use of Theorem II (p. 3) gives $\dfrac{1}{a} \ln \tan \left(\dfrac{\pi}{4} + \dfrac{ax}{2} \right).$

3.2 Use of trigonometric identities

Often in the integration of trigonometric functions we have to use trigonometric identities. Thus for products such as $\sin ax \cos bx$, where $a \neq b$, we proceed as follows:

(i) $\displaystyle\int \sin ax \cos bx\, dx = \int \frac{1}{2} [\sin (a+b)x + \sin (a-b)x]\, dx$

$$= \frac{1}{2} \left[\frac{\cos (a+b)x}{-(a+b)} + \frac{\cos (a-b)x}{-(a-b)} \right],$$

(ii) $\displaystyle\int \cos ax \cos bx\, dx = \int \frac{1}{2} [\cos (a+b)x + \cos (a-b)x]\, dx$

$$= \frac{1}{2} \left[\frac{\sin (a+b)x}{(a+b)} + \frac{\sin (a-b)x}{(a-b)} \right],$$

(iii) $\displaystyle\int \sin ax \sin bx\, dx = \int \frac{1}{2} [\cos (a-b)x - \cos (a+b)x]\, dx$

$$= \frac{1}{2} \left[\frac{\sin (a-b)x}{(a-b)} - \frac{\sin (a+b)x}{(a+b)} \right].$$

For the squares,

$$\int \cos^2 ax\, dx = \int \frac{1}{2}(\cos 2ax + 1)\, dx = \frac{1}{2} \left[\frac{\sin 2ax}{2a} + x \right],$$

$$\int \sin^2 ax\, dx = \int \frac{1}{2}(1 - \cos 2ax)\, dx = \frac{1}{2} \left[x - \frac{\sin 2ax}{2a} \right],$$

$$\int \tan^2 ax\, dx = \int (\sec^2 ax - 1)\, dx = \frac{1}{a} \tan ax - x,$$

$$\int \csc^2 ax\, dx = \int \sec^2 (\pi/2 - ax)\, dx = -\frac{1}{a} \tan (\pi/2 - ax) = -\frac{1}{a} \cot ax,$$

$$\int \cot^2 ax\, dx = \int (\csc^2 ax - 1)\, dx = -\frac{1}{a} \cot ax - x.$$

Exercise 3.2
Integrate: **1** $\sin 2x \cos 2x$, **2** $\sin 4t \sin 6t$, **3** $\cos 7x \cos 5x$, **4** $\cos 5x \sin x$,
5 $\tan^2 2x$, **6** $\sec 2x$, **7** $\operatorname{cosec}^2 (\pi/2 - t)$, **8** $\sin (1 - x) \cos (1 + x)$, **9** $\cot 3t$.

3.3 More elaborate examples
Integrals involving higher powers of the trigonometric functions have to be dealt with individually and it is not possible to lay down any general method other than to be prepared to use the trigonometric identities together with the standard integrals.

Example 1

(i) $$\int \sin^5 x \, dx = \int (\sin^2 x)^2 \sin x \, dx$$

$$= \int (1 - \cos^2 x)^2 \sin x \, dx$$

$$= -\int (1 - 2 \cos^2 x + \cos^4 x) \, d(\cos x)$$

$$= -\left(\cos x - \frac{2}{3} \cos^3 x + \frac{1}{5} \cos^5 x \right),$$

(ii) $$\int \cos^4 2x \, dx = \int (\cos^2 2x)^2 \, dx = \int \frac{(\cos 4x + 1)^2}{4} \, dx$$

$$= \int \left[\frac{1}{4} \cos^2 4x + \frac{1}{2} \cos 4x + \frac{1}{4} \right] dx$$

$$= \int \left[\frac{1}{8} (\cos 8x + 1) + \frac{1}{2} \cos 4x + \frac{1}{4} \right] dx$$

$$= \int \left[\frac{1}{8} \cos 8x + \frac{1}{2} \cos 4x + \frac{3}{8} \right] dx$$

$$= \frac{1}{64} \sin 8x + \frac{1}{8} \sin 4x + \frac{3x}{8},$$

(iii) $$\int \sec^4 x \, dx = \int \sec^2 x \cdot \sec^2 x \, dx$$

$$= \int (1 + \tan^2 x) \, d(\tan x) = \tan x + \frac{1}{3} \tan^3 x,$$

(iv) $$\int \frac{\sin^2 x}{\cos^4 x} \, dx = \int \tan^2 x \sec^2 x \, dx = \int \tan^2 x \, d(\tan x) = \frac{1}{3} \tan^3 x,$$

(v) $\displaystyle\int \frac{\sin^5 x}{\cos^2 x}\,dx = \int \frac{\sin^4 x \sin x}{\cos^2 x}\,dx$

$$= \int \frac{(1 - \cos^2 x)^2}{\cos^2 x}\sin x\,dx$$

$$= -\int \left(\frac{1}{\cos^2 x} - 2 + \cos^2 x\right)d(\cos x)$$

$$= \frac{1}{\cos x} + 2 \cos x - \frac{\cos^3 x}{3}.$$

Integrals of the form $\displaystyle\int \frac{1}{a + b \cos x}\,dx$ or $\displaystyle\int \frac{1}{a + b \sin x}\,dx$ can always be worked out using the tan half-angle substitution, i.e. $t = \tan\dfrac{x}{2}$, together with the trigonometric formulae $\sin x = \dfrac{2t}{1 + t^2}$, $\cos x = \dfrac{1 - t^2}{1 + t^2}$ and $\tan x = \dfrac{2t}{1 - t^2}$.

Example 2

$$\int \frac{1}{5 - 3 \cos x}\,dx.$$

Let $t = \tan\dfrac{x}{2}$

$$\Rightarrow \frac{dt}{dx} = \frac{1}{2}\sec^2\left(\frac{x}{2}\right) = \frac{1}{2}\left[1 + \tan^2\left(\frac{x}{2}\right)\right] \Rightarrow \frac{dx}{dt} = \frac{2}{1 + t^2}.$$

$$\int \frac{1}{5 - 3 \cos x}\,dx = \int \frac{1}{5 - 3\dfrac{(1 - t^2)}{(1 + t^2)}}\cdot\frac{2}{1 + t^2}\,dt.$$

$$= \int \frac{2}{5(1 + t^2) - 3(1 - t^2)}\,dt$$

$$= \int \frac{1}{1 + 4t^2}\,dt$$

$$= \frac{1}{4}\int \frac{1}{\frac{1}{4} + t^2}\,dt$$

$$= \frac{1}{4}\frac{1}{(\frac{1}{2})}\tan^{-1}\left(\frac{t}{(\frac{1}{2})}\right)$$

$$= \frac{1}{2}\tan^{-1} 2t$$

$$= \frac{1}{2}\tan^{-1}\left(2 \tan\frac{x}{2}\right).$$

Note, however, that integrals involving functions of $\sin^2 x$ or $\sin x \cos x$ or $\cos^2 x$ may be determined by using the substitution $t = \tan x$.

Example 3

$$\int \frac{1}{1 + \sin^2 x} dx.$$

Let $t = \tan x$

$$\Rightarrow \frac{dt}{dx} = \sec^2 x = 1 + \tan^2 x = 1 + t^2 \Rightarrow \frac{dx}{dt} = \frac{1}{1 + t^2}.$$

$$\int \frac{1}{1 + \sin^2 x} dx = \int \frac{1}{\cos^2 x (\sec^2 x + \tan^2 x)} dx$$

$$= \int \frac{\sec^2 x}{1 + 2 \tan^2 x} dx$$

$$= \int \frac{(1 + t^2)}{1 + 2t^2} \cdot \frac{1}{1 + t^2} dt$$

$$= \int \frac{1}{1 + 2t^2} dt$$

$$= \frac{1}{2} \int \frac{1}{\frac{1}{2} + t^2} dt$$

$$= \frac{1}{2} \frac{1}{\sqrt{(\frac{1}{2})}} \tan^{-1} \left[\frac{t}{(1/\sqrt{2})} \right]$$

$$= \frac{1}{\sqrt{2}} \tan^{-1} (t\sqrt{2})$$

$$= \frac{1}{\sqrt{2}} \tan^{-1} [\sqrt{2} \tan x].$$

Exercise 3.3

Integrate: **1** $\sin^3 x$, **2** $\sin^4 x$, **3** $\sin^3 x \cos^2 x$, **4** $\sec^6 x$, **5** $\dfrac{\sin^4 x}{\cos^6 x}$, **6** $\dfrac{\sin^5 x}{\cos^6 x}$,

7 $\tan^4 x$, **8** $\operatorname{cosec}^4 x$, **9** $\dfrac{1}{1 - \cos x}$, **10** $\dfrac{1}{1 + \sin x}$, **11** $\dfrac{1}{1 + 15 \cos^2 x}$,

12 $\dfrac{1}{1 + \sin x \cos x}$.

4 Integration of rational algebraic fractions

4.1 Introduction

Functions of the form $\dfrac{a_0 + a_1 x + a_2 x^2 + \ldots + a_n x^n}{b_0 + b_1 x + b_2 x^2 + \ldots + b_m x^m}$, where m and n are positive integers and $a_0, a_1, \ldots, a_n, b_0, b_1, \ldots, b_m$ are all constants, are known as rational algebraic fractions. The integration of such functions for small values of m and n can be very carefully categorised.

If n, the degree of the numerator, is greater than or equal to m, the degree of the denominator, then, before any integration can be considered, the numerator must be divided by the denominator until the remainder is of a lower degree than the denominator. We are left with the remainder

$$\frac{c_0 + c_1 x + c_2 x^2 + \ldots + c_{m-1} x^{m-1}}{b_0 + b_1 x + b_2 x^2 + \ldots + b_m x^m}$$

which can then be considered to fall into certain categories.

4.2 The denominator is of the first degree

This means the remainder is of the form $\dfrac{c_0}{b_0 + b_1 x}$, the integral of which is

$$\int \frac{c_0}{b_0 + b_1 x} dx = \frac{c_0}{b_1} \ln (b_0 + b_1 x).$$

Thus, for $\displaystyle\int \frac{x^3}{x - 1} dx$, division gives

$$\int \frac{x^3}{x - 1} dx = \int \left[x^2 + x + 1 + \frac{1}{x - 1} \right] dx = \frac{x^3}{3} + \frac{x^2}{2} + x + \ln (x - 1).$$

Similarly,

$$\int \frac{4x + 5}{2 - 3x} dx = \int \left[-\frac{4}{3} + \frac{23}{3} \cdot \frac{1}{2 - 3x} \right] dx$$

$$= -\frac{4x}{3} - \frac{23}{9} \ln (2 - 3x).$$

Exercise 4.2

Integrate: **1** $\dfrac{2x}{x + 4}$, **2** $\dfrac{x^2}{3 - x}$, **3** $\dfrac{2x + 1}{1 - 3x}$, **4** $\dfrac{3x^2 + 6}{3x + 2}$.

4.3 The denominator is of the second degree and breaks up into rational factors

Such integrals are solved by breaking up the expression into partial fractions, thus reducing it to the type of integral in §4.2.

Example 1

(i) $\displaystyle\int \frac{2x + 4}{2x^2 - x - 1}\,dx = \int \frac{2x + 4}{(x - 1)(2x + 1)}\,dx$

$$= \int\left[\frac{2}{x - 1} - \frac{2}{2x + 1}\right]dx$$

$$= 2\ln(x - 1) - \ln(2x + 1),$$

(ii) $\displaystyle\int \frac{2x^2}{x^2 - 1}\,dx = \int\left(2 + \frac{2}{x^2 - 1}\right)dx$

$$= \int\left(2 - \frac{1}{x + 1} + \frac{1}{x - 1}\right)dx$$

$$= 2x - \ln(x + 1) + \ln(x - 1),$$

(iii) $\displaystyle\int \frac{x + 3}{x^2 + 4x + 4}\,dx = \int \frac{x + 3}{(x + 2)^2}\,dx$

$$= \int\left(\frac{1}{x + 2} + \frac{1}{(x + 2)^2}\right)dx$$

$$= \ln(x + 2) - \frac{1}{(x + 2)}.$$

Exercise 4.3

Integrate: **1** $\dfrac{5}{x^2 - x - 6}$, **2** $\dfrac{2x + 3}{x^2 + x - 20}$, **3** $\dfrac{x^2 + 1}{1 - 4x^2}$, **4** $\dfrac{x + 2}{(x - 1)^2}$, **5** $\dfrac{3x + 2}{5x^2 + 3x}$,

6 $\dfrac{x^3}{9 - x^2}$, **7** $\dfrac{4x + 3}{3x^2 + 10x + 3}$.

The results of two integrals of the kind considered in §4.3 should be noted as they are particularly useful in the solution of the type of §4.4. They are

$$\int \frac{1}{x^2 - a^2}\,dx = \int \frac{1}{2a}\left(\frac{1}{x - a} - \frac{1}{x + a}\right)dx, \quad a > 0,$$

$$= \frac{1}{2a}[\ln(x - a) - \ln(x + a)]$$

$$= \frac{1}{2a}\ln\left(\frac{x - a}{x + a}\right), \quad \text{for } |x| > a.$$

and

$$\int \frac{1}{a^2 - x^2}\,dx = \int \frac{1}{2a}\left(\frac{1}{a+x} + \frac{1}{a-x}\right)dx$$

$$= \frac{1}{2a}[\ln(a+x) - \ln(a-x)]$$

$$= \frac{1}{2a}\ln\left(\frac{a+x}{a-x}\right) \text{ for } |x| < a.$$

4.4 The denominator is of the second degree, it does not break up into rational factors and the numerator is a constant

This type can be reduced to one of the two above forms or to the form

$$\int \frac{1}{a^2 + x^2}\,dx = \frac{1}{a}\tan^{-1}\left(\frac{x}{a}\right).$$

Example 2

(i) $\displaystyle\int \frac{1}{x^2 + 2x + 5}\,dx = \int \frac{1}{(x+1)^2 + 2^2}\,dx$

$$= \frac{1}{2}\tan^{-1}\left(\frac{x+1}{2}\right), \quad \text{by Theorem I (p. 2).}$$

(ii) $\displaystyle\int \frac{1}{x^2 + 6x + 4}\,dx = \int \frac{1}{(x+3)^2 - 5}\,dx$

and, comparing with the results for $\displaystyle\int \frac{1}{x^2 - a^2}\,dx$, we have

for $|x+3| > \sqrt{5}$, $\displaystyle\int \frac{1}{x^2 + 6x + 4}\,dx = \frac{1}{2\sqrt{5}}\ln\left(\frac{x+3-\sqrt{5}}{x+3+\sqrt{5}}\right)$

or, for $|x+3| < \sqrt{5}$, $\displaystyle\int \frac{1}{x^2 + 6x + 4} = \frac{1}{2\sqrt{5}}\ln\left(\frac{\sqrt{5}+(x+3)}{\sqrt{5}-(x+3)}\right).$

(iii) $\displaystyle\int \frac{3}{3 + 4x - 2x^2}\,dx = \frac{1}{2}\int \frac{3}{\frac{3}{2} + 2x - x^2}\,dx = \frac{1}{2}\int \frac{3}{\frac{5}{2} - (x-1)^2}\,dx,$

which, comparing with $\displaystyle\int \frac{1}{a^2 - x^2}\,dx$ for $|x| \lessdot a$, gives for $|x - 1| < \sqrt{(5/2)}$:

$$\int \frac{3}{3 + 4x - 2x^2}\,dx = \frac{1}{2}\cdot\frac{1}{2\sqrt{(5/2)}}\ln\left(\frac{\sqrt{(5/2)}+(x-1)}{\sqrt{(5/2)}-(x-1)}\right)$$

$$= \frac{1}{2\sqrt{10}}\ln\left(\frac{\sqrt{5}+\sqrt{2}(x-1)}{\sqrt{5}-\sqrt{2}(x-1)}\right).$$

Exercise 4.4

Integrate: **1** $\dfrac{1}{x^2 + 4x + 5}$, **2** $\dfrac{1}{2x^2 + 3x + 2}$, **3** $\dfrac{3}{4 - 3x^2}$, **4** $\dfrac{2}{4 - 2x - x^2}$.

4.5 The denominator is of the second degree, it does not break up into rational factors, the numerator is not constant but is first degree

Integrals of this type are solved by splitting them up into two parts – one of the form $\displaystyle\int \dfrac{\mathrm{d}f(x)}{f(x)}$, the other of the type of §4.4.

Example 3

(i) $\displaystyle\int \dfrac{3x + 1}{x^2 + 2x + 5}\,\mathrm{d}x$.

Differentiating the denominator to obtain $\mathrm{d}f(x)$ gives

$$\int \frac{3x + 1}{x^2 + 2x + 5}\,\mathrm{d}x = \int \frac{\frac{3}{2}(2x + 2) - 2}{x^2 + 2x + 5}\,\mathrm{d}x$$

$$= \frac{3}{2}\int \frac{(2x + 2)}{x^2 + 2x + 5}\,\mathrm{d}x - 2\int \frac{1}{(x + 1)^2 + 4}\,\mathrm{d}x$$

$$= \frac{3}{2}\ln(x^2 + 2x + 5) - \frac{2}{2}\tan^{-1}\left(\frac{x + 1}{2}\right).$$

(ii) $\displaystyle\int \dfrac{x^2 + 4x + 2}{4 - 2x - x^2}\,\mathrm{d}x$

$$= \int \left(-1 + \frac{2x + 6}{4 - 2x - x^2}\right)\mathrm{d}x$$

$$= \int \left(-1 - \frac{(-2x - 2) - 4}{4 - 2x - x^2}\right)\mathrm{d}x$$

$$= \int \left(-1 - \frac{-2x - 2}{4 - 2x - x^2} + \frac{4}{5 - (1 + x)^2}\right)\mathrm{d}x$$

$$= -x - \ln(4 - 2x - x^2) + 4 \cdot \frac{1}{2\sqrt{5}}\ln\left(\frac{\sqrt{5} + (1 + x)}{\sqrt{5} - (1 - x)}\right).$$

Exercise 4.5

Integrate: **1** $\dfrac{3x + 2}{x^2 - 5}$, **2** $\dfrac{4x + 1}{2x^2 + 2x + 1}$, **3** $\dfrac{x^2 - 1}{x^2 - 2x + 5}$, **4** $\dfrac{5 + 2x}{1 + 4x - 2x^2}$.

4.6 The denominator is of a higher degree than the second

When the denominator is of a higher degree than the second, one must try to factorise the denominator and trust that by the use of partial fractions the expression may be integrated by means of one of the previous methods.

Example 4

(i) $\displaystyle\int \frac{1}{x^2(x+1)}\,dx = \int \left(-\frac{1}{x} + \frac{1}{x^2} + \frac{1}{x+1}\right)dx$

$\qquad\qquad\qquad = -\ln x - \frac{1}{x} + \ln(x+1).$

(ii) $\displaystyle\int \frac{5x^2}{(x-1)(x^2+4)}\,dx = \int \left(\frac{1}{x-1} + \frac{4x+4}{x^2+4}\right)dx$

$\qquad\qquad\qquad\qquad = \int \left(\frac{1}{x-1} + 2\cdot\frac{2x}{x^2+4} + 4\cdot\frac{1}{x^2+4}\right)dx$

$\qquad\qquad\qquad\qquad = \ln(x-1) + 2\ln(x^2+4) + 4\cdot\tfrac{1}{2}\tan^{-1}(x/2).$

Exercise 4.6

Integrate: **1** $\dfrac{2}{x(x^2-1)}$, **2** $\dfrac{1}{(x+2)(x^2+1)}$, **3** $\dfrac{x}{(x+1)(x^2+2x+2)}$, **4** $\dfrac{1}{x^3-8}$.

5 Integration of irrational algebraic fractions of the form $\dfrac{ax + b}{\sqrt{(cx^2 + dx + e)}}$, $c < 0$

5.1 Case I: $a = 0$

If $a = 0$, then the integral can be reduced to $\displaystyle\int \frac{1}{\sqrt{(a^2 - x^2)}}\,dx = \sin^{-1}\left(\frac{x}{a}\right)$.

Example 1

(i) $\displaystyle\int \frac{1}{\sqrt{(8 + 2x - x^2)}}\,dx = \int \frac{1}{\sqrt{[9 - (x - 1)^2]}}\,dx = \sin^{-1}\left(\frac{x - 1}{3}\right)$.

(ii) $\displaystyle\int \frac{1}{\sqrt{(1 - 8x - 4x^2)}}\,dx = \frac{1}{2}\int \frac{1}{\sqrt{(\frac{1}{4} - 2x - x^2)}}\,dx$

$\qquad\qquad = \dfrac{1}{2}\displaystyle\int \frac{1}{\sqrt{[\frac{5}{4} - (x + 1)^2]}}\,dx$

$\qquad\qquad = \dfrac{1}{2}\sin^{-1}\left(\dfrac{x + 1}{\sqrt{5/2}}\right)$

$\qquad\qquad = \dfrac{1}{2}\sin^{-1}\left(\dfrac{2(x + 1)}{\sqrt{5}}\right)$.

5.2 Case II: $a \neq 0$, $c < 0$

If $a \neq 0$, we make use of the result $\displaystyle\int \frac{\frac{1}{2}\frac{du}{dx}}{\sqrt{u}}\,dx = \sqrt{u}$, where $u = f(x)$.

Hence, to integrate expressions such as $\dfrac{ax + b}{\sqrt{(cx^2 + dx + e)}}$ we arrange the numerator in the form $k\dfrac{1}{2}\cdot\dfrac{d}{dx}(cx^2 + dx + e) + l$, where k and l are constants. This gives an integral $k\sqrt{(cx^2 + dx + e)}$ together with one of Case 1.

Example 2

(i) $\displaystyle\int \frac{3x + 2}{\sqrt{(3 + 2x - x^2)}}\,dx = \int \frac{-3\cdot\frac{1}{2}(2 - 2x) + 5}{\sqrt{(3 + 2x - x^2)}}\,dx$

$\qquad\qquad = -3\displaystyle\int \frac{\frac{1}{2}(2 - 2x)}{\sqrt{(3 + 2x - x^2)}}\,dx + \int \frac{5}{\sqrt{[4 - (x - 1)^2]}}\,dx$

$\qquad\qquad = -3\sqrt{(3 + 2x - x^2)} + 5\sin^{-1}\left(\dfrac{x - 1}{2}\right)$.

(ii) $\displaystyle\int \sqrt{\left(\frac{4-x}{x}\right)}\,dx = \int \frac{4-x}{\sqrt{[x(4-x)]}}\,dx$

$\displaystyle\qquad = \int \frac{\frac{1}{2}(4-2x)+2}{\sqrt{(4x-x^2)}}\,dx$

$\displaystyle\qquad = \int \frac{\frac{1}{2}(4-2x)}{\sqrt{(4x-x^2)}}\,dx + 2\int \frac{1}{\sqrt{[4-(x-2)^2]}}\,dx$

$\displaystyle\qquad = \sqrt{(4x-x^2)} + 2\sin^{-1}\left(\frac{x-2}{2}\right).$

Exercise 5.2

Integrate: **1** $\dfrac{x+1}{\sqrt{(4-x^2)}}$, **2** $\dfrac{2x+3}{\sqrt{(5+4x-x^2)}}$, **3** $\sqrt{\left(\dfrac{4-x}{2+x}\right)}.$

6 Integration by parts

6.1 Introduction

We are sometimes faced with either an integrand which we would find easier to differentiate than to integrate or an integral of a product which is such that one part of it can be integrated easily but the other part cannot. Examples of such integrals include $\int \ln x\, dx$ or $\int x \sin^{-1} x\, dx$. In such circumstances it is worth trying a method known as 'integration by parts' which is obtained from the derivative of a product.

$$\frac{d}{dx}(u.v) = u\frac{dv}{dx} + v\frac{du}{dx} \;\Rightarrow\; uv = \int \left(u\frac{dv}{dx} + v\frac{du}{dx}\right) dx$$

and, rearranging, we obtain

$$\int u\frac{dv}{dx}dx = uv - \int v\frac{du}{dx}dx,$$

where we note that of the two terms on the left hand side one has to be integrated, i.e. $\frac{dv}{dx}$, while the other, u, has to be differentiated.

Example 1 If one of the terms is very difficult to integrate then obviously choose this to be u, the one that is to be differentiated. For example

$$\int x^3 \ln x\, dx = \frac{x^4}{4}\cdot \ln x - \int \frac{x^4}{4}\cdot\frac{1}{x}dx$$

$$\begin{array}{ccccc} \frac{dv}{dx} & u & v & u & v & \frac{du}{dx} \end{array}$$

$$= \frac{x^4}{4}\ln x - \frac{x^4}{16}.$$

Example 2 If both parts of the integrand are easy to integrate separately, then choose the one to be differentiated as the one that will ultimately give zero if differentiated a sufficient number of times. For example $\int x^2 e^{3x}\, dx$.

If we differentiate x^2 three times we obtain zero. No matter how many times we differentiate e^{3x} we never get zero.

Choosing $$x^2 \equiv u \quad \text{and} \quad e^{3x} \equiv \frac{dv}{dx}$$

$$\Rightarrow \int x^2 e^{3x} \, dx = x^2 \cdot \frac{e^{3x}}{3} - \int \frac{e^{3x}}{3} \cdot 2x \, dx$$

$$\qquad\qquad u\frac{dv}{dx} \qquad\quad u \quad v \qquad\quad v \quad \frac{du}{dx}$$

and, repeating the process,

$$\int x^2 e^{3x} \, dx = \frac{1}{3} x^2 e^{3x} - \frac{2}{3} \left[\frac{e^{3x}}{3} \cdot x - \int \frac{e^{3x}}{3} \cdot 1 \, dx \right]$$

$$= \frac{1}{3} x^2 e^{3x} - \frac{2}{9} x e^{3x} + \frac{2}{27} e^{3x}.$$

Had we chosen the terms the other way round, i.e. $e^{3x} \equiv u$ and $x^2 \equiv \dfrac{dv}{dx}$, then we should not have succeeded in determining the integral but instead have been left with a more difficult integral.

Example 3 If both parts of the integrand are easy to integrate separately and neither gives zero on repeated differentiation, then it does not matter which way we choose u and $\dfrac{dv}{dx}$. Either way we are faced with a repeated application of the method. For example,

$$\int e^{2x} \cos 3x \, dx = e^{2x} \cdot \frac{\sin 3x}{3} - \int \frac{\sin 3x}{3} \cdot 2e^{2x} \, dx$$

$$\quad u \qquad\qquad \frac{dv}{dx} \qquad u \qquad v \qquad\qquad v \qquad \frac{du}{dx}$$

$$= \frac{1}{3} e^{2x} \sin 3x - \frac{2}{3} \int e^{2x} \sin 3x \, dx.$$

Repeating,

$$\int e^{2x} \cos 3x \, dx = \frac{1}{3} e^{2x} \sin 3x - \frac{2}{3} \left[e^{2x} \left(\frac{\cos 3x}{-3} \right) - \int \frac{\cos 3x}{-3} \cdot 2e^{2x} \, dx \right]$$

$$= \frac{1}{3} e^{2x} \sin 3x + \frac{2}{9} e^{2x} \cos 3x - \frac{4}{9} \int e^{2x} \cos 3x \, dx$$

$$\Rightarrow \left(1 + \frac{4}{9} \right) \int e^{2x} \cos 3x \, dx = \frac{1}{3} e^{2x} \sin 3x + \frac{2}{9} e^{2x} \cos 3x$$

$$\Rightarrow \int e^{2x} \cos 3x \, dx = \frac{e^{2x}}{13} (3 \sin 3x + 2 \cos 3x).$$

Example 4 The method is sometimes applicable to functions which are not products but which are easier to differentiate than integrate, in particular a logarithmic function or an inverse trigonometric function. For example

$$\int \underset{\substack{\\ \dfrac{dv}{dx}}}{\tan^{-1} x}\, \underset{u}{} \, dx = \int \underset{v}{1}\,.\, \underset{u}{\tan^{-1} x}\, dx = x\tan^{-1} x - \int \underset{v}{x}\cdot \underset{\dfrac{du}{dx}}{\dfrac{1}{1+x^2}}\, dx$$

$$= x\tan^{-1} x - \tfrac{1}{2}\ln (1 + x^2).$$

Similarly,

$$\int \ln x\, dx = x\ln x - \int x\cdot\frac{1}{x}\,dx = x\ln x - x.$$

Exercise 6.1

Integrate: **1** $x \sin 3x$, **2** $\dfrac{\ln x}{x^2}$, **3** $x^2 e^{4x}$, **4** $\ln (2 + x)$, **5** $\sin^{-1} x$, **6** $e^x \cos 2x$, **7** $\sqrt{(1 - x^2)}$, **8** $(\ln x)^2$, **9** $x\sqrt{(1 + x)}$, **10** $x \sin^2 x$.

7 Definite integrals

7.1 Definitions

The integral of a function of x is itself a new function of x. The applications of integration, which we shall later consider, often require the values of the new function for different values of x. In particular we often require the difference in value of the new function for two particular values of x, say $x = a$ and $x = b$. This we write as

$$\int_a^b f(x)\,dx = \left[F(x) \right]_a^b = F(b) - F(a),$$

and we call this the definite integral of $f(x)$ with respect to x over the range $x = a$ to $x = b$, or alternatively between the limits $x = a$ (the lower limit) and $x = b$ (the upper limit).

Example 1

(i) $\displaystyle \int_1^2 \left(x^2 + \frac{1}{x^2} \right) dx = \left[\frac{x^3}{3} - \frac{1}{x} \right]_1^2 = \left(\frac{8}{3} - \frac{1}{2} \right) - \left(\frac{1}{3} - 1 \right) = 2\frac{5}{6},$

(ii) $\displaystyle \int_0^{\pi/2} \sin^3 x \cos x\,dx = \int_0^{\pi/2} \sin^3 x\,d(\sin x) = \left[\frac{\sin^4 x}{4} \right]_0^{\pi/2}$

$$= \frac{1}{4} \sin^4 (\pi/2) - \sin^4 0 = \frac{1}{4},$$

(iii) $\displaystyle \int_a^b \frac{1}{x+1}\,dx = \left[\ln (x+1) \right]_a^b = \ln(b+1) - \ln(a+1) = \ln \left(\frac{b+1}{a+1} \right).$

7.2 Substitution methods involving limits

If it is found necessary to make a substitution in order to evaluate a definite integral, then one should change the limits during integration so as to accommodate the new variable. Note that, if we make the substitution $x = \phi(t)$ in

$$I = \int_a^b f(x)\,dx, \text{ then } I = \int_{t_0}^{t_1} f[\phi(t)]\phi'(t)\,dt, \text{ where } a = \phi(t_0),\ b = \phi(t_1), \text{ only if}$$

$\phi'(t)$ has constant sign between t_0 and t_1.

Alternatively one can treat the integral as an indefinite integral (i.e. one without limits) and insert the original limits at the end of the integration after the function has been expressed in terms of the original variable. This, however, is usually more involved and is a method not really to be recommended.

Example 2

$$\int_0^2 \sqrt{(4 - x^2)}\,dx.$$

Let $x = 2 \sin \theta \Rightarrow \dfrac{dx}{d\theta} = 2 \cos \theta.$

When $x = 2$, $\sin \theta = 1 \Rightarrow \theta = \pi/2$; when $x = 0$, $\sin \theta = 0 \Rightarrow \theta = 0$

$$\Rightarrow \int_0^2 \sqrt{(4 - x^2)}\,dx = \int_0^{\pi/2} 4 \cos^2 \theta\,d\theta$$

$$= \int_0^{\pi/2} 2(1 + \cos 2\theta)\,d\theta = 2\left[\theta + \tfrac{1}{2} \sin 2\theta \right]_0^{\pi/2}$$

$$= 2 \cdot \frac{\pi}{2} + \sin \pi - 0 = \pi.$$

Alternatively,

$$2\left[\theta + \tfrac{1}{2} \sin 2\theta\right] = 2[\theta + \sin \theta \cos \theta] = 2\left[\sin^{-1}\left(\frac{x}{2}\right) + \tfrac{1}{4}x\sqrt{(4 - x^2)} \right],$$

$$\Rightarrow \int_0^2 \sqrt{(4 - x^2)}\,dx = 2\left[\sin^{-1}\left(\frac{x}{2}\right) + \tfrac{1}{4}x\sqrt{(4 - x^2)} \right]_0^2$$

$$= 2[\sin^{-1} 1 + 0] - 2[\sin^{-1} 0 + 0] = 2(\pi/2) = \pi.$$

Example 3

$$\int_0^1 (1 - t)^{3/2}t\,dt.$$

Let $t = \sin^2 \theta \Rightarrow \dfrac{dt}{d\theta} = 2 \sin \theta \cos \theta.$

When $t = 1$, $\sin^2 \theta = 1 \Rightarrow \theta = \pi/2$; when $t = 0$, $\sin^2 \theta = 0 \Rightarrow \theta = 0$

$$\Rightarrow \int_0^1 (1 - t)^{3/2}t\,dt = \int_0^{\pi/2} (1 - \sin^2 \theta)^{3/2} \sin^2 \theta \,.\, 2 \sin \theta \cos \theta\,d\theta$$

$$= 2 \int_0^{\pi/2} \cos^4 \theta \sin^3 \theta\,d\theta$$

$$= -2 \int_0^{\pi/2} \cos^4 \theta(1 - \cos^2 \theta)\,d(\cos \theta)$$

$$= -2\left[\frac{\cos^5 \theta}{5} - \frac{\cos^7 \theta}{7} \right]_0^{\pi/2}$$

$$= 2\left(\frac{1}{5} - \frac{1}{7} \right) = \frac{4}{35}.$$

Example 4

$$\int_{2}^{2\sqrt{3}} \frac{1}{x^2\sqrt{(4 + x^2)}}\,dx.$$

Let $x = 2\tan\theta \Rightarrow \dfrac{dx}{d\theta} = 2\sec^2\theta.$

When $x = 2\sqrt{3},\quad \tan\theta = \sqrt{3} \Rightarrow \theta = \pi/3,$
When $x = 2,\quad \tan\theta = 1 \Rightarrow \theta = \pi/4,$

$$\Rightarrow \int_{\pi/4}^{\pi/3} \frac{2\sec^2\theta}{4\tan^2\theta\sqrt{(4 + 4\tan^2\theta)}}\,d\theta = \int_{\pi/4}^{\pi/3} \frac{2\sec^2\theta}{8\tan^2\theta\sec\theta}\,d\theta$$

$$= \frac{1}{4}\int_{\pi/4}^{\pi/3} \frac{\cos\theta}{\sin^2\theta}\,d\theta$$

$$= \frac{1}{4}\int_{\pi/4}^{\pi/3} \frac{1}{\sin^2\theta}\,d(\sin\theta)$$

$$= \frac{1}{4}\left[-\frac{1}{\sin\theta}\right]_{\pi/4}^{\pi/3} = \frac{1}{4}\left[-\frac{2}{\sqrt{3}} + \sqrt{2}\right].$$

Exercise 7.2

Evaluate: **1** $\displaystyle\int_{0}^{2} x^3(2 - x)^{1/2}\,dx,$ **2** $\displaystyle\int_{0}^{4} \frac{x}{1 + \sqrt{x}}\,dx,$ **3** $\displaystyle\int_{0}^{\pi/4} \frac{(1 + \cos x)\sin x}{\sqrt{(\cos x)}}\,dx,$

4 $\displaystyle\int_{0}^{1} x\sqrt{(1 + x)}\,dx,$ **5** $\displaystyle\int_{0}^{\pi/2} \sin 2\theta\sqrt{(\sin\theta)}\,d\theta,$ **6** $\displaystyle\int_{0}^{1} x(2 + e^{-x^2})\,dx,$

7 $\displaystyle\int_{0}^{\pi/3} \sin 3x \cos 2x\,dx,$ **8** $\displaystyle\int_{0}^{\pi/2} \sin^3 t \cos^2 t\,dt,$ **9** $\displaystyle\int_{1/2}^{1} x^2\sqrt{(1 - x^2)}\,dx.$

10 Show, by means of a suitable substitution, that

$$\int_{0}^{\pi/2} \frac{\sin x}{\sin x + \cos x}\,dx = \int_{0}^{\pi/2} \frac{\cos x}{\sin x + \cos x}\,dx$$

and hence determine their value. (AEB)

11 By using suitable substitutions, or otherwise, evaluate

(i) $\displaystyle\int_{1}^{2} x\sqrt{(x - 1)}\,dx,$ (ii) $\displaystyle\int_{0}^{1} \frac{1}{(1 + x^2)^2}\,dx,$ (iii) $\displaystyle\int_{0}^{2} x^3(4 - x^2)^{3/2}\,dx.$ (AEB)

12 Given that $y = \tan^{-1} 3x$, prove that $\dfrac{dy}{dx} = \dfrac{3}{1 + 9x^2}.$ Hence, or otherwise, evaluate

$$\int_{0}^{1} x^2 \tan^{-1} 3x\,dx.$$ (AEB)

8 Applications of integration

8.1 Introduction

The applications of integration are numerous and at an elementary level are perhaps best shown by taking examples. It will be seen that the indefinite integral gives a general solution to a problem and that particular solutions satisfying given initial conditions are obtained by determining specific values of the constant or constants of integration.

Example 1   The gradient at any point (x, y) on a curve is given by $3x + 2$. Find the equation of the curve, given that it passes through the point $(0, 2)$.

The gradient of a curve $y = f(x)$ is given by $\dfrac{dy}{dx}$ and so $\dfrac{dy}{dx} = 3x + 2$.

Integrating, $y = \dfrac{3x^2}{2} + 2x + C$, where C is an arbitrary constant of integration.

This is the equation of all curves passing through the point (x, y) and having a gradient $3x + 2$ at that point. To find the particular curve passing through the point $(0, 2)$, it is now necessary to determine the specific value of the constant of integration C by substituting the values $x = 0$, $y = 2$

$$\Rightarrow 2 = 0 + 0 + C \Rightarrow C = 2$$

$$\Rightarrow y = \frac{3x^2}{2} + 2x + 2.$$

Example 2   A particle moves along a straight line with acceleration $(3 + 2t)\,\mathrm{m\,s^{-2}}$ at time t seconds. Initially the particle is at rest at a distance 5 m from the origin. Find the velocity and distance of the particle from the origin at (i) time t seconds, (ii) time 4 seconds.

Acceleration = rate of change of velocity

$$= \frac{dv}{dt}, \quad \text{where } v\,\mathrm{m\,s^{-1}} \text{ is the velocity at time } t \text{ seconds,}$$

$$\Rightarrow \frac{dv}{dt} = 3 + 2t$$

$$\Rightarrow v = \int (3 + 2t)\,dt = 3t + t^2 + C, \quad \text{where } C \text{ is a constant of integration.}$$

When $t = 0$, $v = 0 \Rightarrow C = 0$ and $v = 3t + t^2$.

Velocity = rate of change of distance

$$= \frac{ds}{dt}, \quad \text{where } s \text{ m is the distance from the origin at time } t \text{ seconds,}$$

$$\Rightarrow v = \frac{ds}{dt} = 3t + t^2$$

$$\Rightarrow s = \int (3t + t^2)\,dt = \frac{3t^2}{2} + \frac{t^3}{3} + K, \quad \text{where } K \text{ is a constant of integration.}$$

When $t = 0, s = 5 \Rightarrow 5 = 0 + 0 + K \Rightarrow s = \dfrac{3t^2}{2} + \dfrac{t^3}{3} + 5.$

At time t seconds: velocity $= (3t + t^2)\,\text{m}\,\text{s}^{-1}$,

$$\text{distance from origin} = \left(\frac{3t^2}{2} + \frac{t^3}{3} + 5 \right)\text{m.}$$

At time 4 seconds: velocity $= (3.4 + 4^2)\,\text{m}\,\text{s}^{-1} = 28\,\text{m}\,\text{s}^{-1}$,

$$\text{distance from origin} = \left(\frac{3}{2}.4^2 + \frac{4^3}{3} + 5 \right)\text{m} = 50\frac{1}{3}\,\text{m.}$$

Example 3   A particle moves in a straight line so that its acceleration at any instant is proportional to the square of the time for which it has been moving. If the particle starts from a fixed point P with initial velocity $u\,\text{m}\,\text{s}^{-1}$, find its velocity and its distance from P after t seconds.

$$\text{Acceleration} \propto t^2 \Rightarrow \text{acceleration} = kt^2, \quad \text{where } k \text{ is a constant,}$$

$$\Rightarrow \frac{dv}{dt} = kt^2$$

$$\Rightarrow v = \int kt^2\,dt = \frac{kt^3}{3} + A, \quad \text{where } A \text{ is a constant of integration.}$$

When $t = 0, v = u \Rightarrow A = u,$

$$\Rightarrow \text{velocity at time } t \text{ seconds is} \quad \frac{kt^3}{3} + u.$$

Let x be the distance of the particle from P at time t seconds. Then $\dfrac{dx}{dt} = \dfrac{1}{3}kt^3 + u \Rightarrow x = \int \left(\dfrac{1}{3}kt^3 + u \right)dt = \dfrac{1}{12}kt^4 + ut + B$, where B is a constant of integration.

When $t = 0, x = 0 \Rightarrow B = 0 \Rightarrow \text{distance from } P \text{ after } t \text{ seconds} = \dfrac{1}{12}kt^4 + ut.$

Example 4   If $\dfrac{d^2y}{dx^2} = 3x^2 - 2x + 4$, find y in terms of x given that, when

$x = 1, y = -1$ and $\dfrac{dy}{dx} = 2$.

$$\dfrac{d^2y}{dx^2} = 3x^2 - 2x + 4 \Rightarrow \dfrac{dy}{dx} = \int(3x^2 - 2x + 4)\,dx = x^3 - x^2 + 4x + C.$$

When $x = 1, \dfrac{dy}{dx} = 2 \Rightarrow 2 = 1 - 1 + 4 + C \Rightarrow C = -2$

$$\Rightarrow \dfrac{dy}{dx} = x^3 - x^2 + 4x - 2$$

$$\Rightarrow \quad y = \int(x^3 - x^2 + 4x - 2)\,dx = \dfrac{x^4}{4} - \dfrac{x^3}{3} + 2x^2 - 2x + K.$$

When $x = 1, y = -1 \Rightarrow -1 = \dfrac{1}{4} - \dfrac{1}{3} + 2 - 2 + K \Rightarrow K = -\dfrac{11}{12}$

$$\Rightarrow y = \dfrac{x^4}{4} - \dfrac{x^3}{3} + 2x^2 - 2x - \dfrac{11}{12}.$$

Exercise 8.1

1   Find the equation of the curves in which the slope is proportional to the square root of the abscissa. [The abscissa is the x-coordinate.]

2   A curve passes through the origin of coordinates and its gradient at the point (x, y) is $3x - \dfrac{x^2}{2}$. Find the ordinate of the curve when $x = 2$. [The ordinate is the y-coordinate.]

3   The gradient of the normal to a curve at any point is inversely proportional to the abscissa of that point. Find the equation of the curve given that it passes through the points $(1, 1)$ and $(3, 13)$.

4   Find the equation of the family of curves for which the subtangent is of constant length c.

5   A particle starting with velocity of $15\ \text{m s}^{-1}$ has an acceleration $(1 - 4t)\ \text{m s}^{-2}$. Find the time taken before it comes to rest and the distance it has then travelled.

6   If $\dfrac{d^2y}{dx^2} = 7 + 2x - 6\sin 3x$, find y in terms of x given that, when $x = 0, \dfrac{dy}{dx} = 1$ and $y = 2$.

7   If $\dfrac{d^2y}{dx^2} = \dfrac{1}{(2 - x)^3}$, find y in terms of x given that, when $x = 1, y = 2$, and when $x = 0, y = -\frac{1}{4}$.

8.2   Area under a curve

Consider the area of the finite region $BCQP$ contained between the arc PQ of the curve with equation $y = f(x)$, the x-axis and the ordinates BP and CQ at $x = b$ and $x = c$ respectively, as shown in Fig. 8.1.

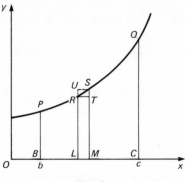

Fig. 8.1

Suppose the area is divided into n strips of equal width δx of which $LMSR$ is one such strip.

If $R \equiv (x, y)$ and $S \equiv (x + \delta x, y + \delta y)$, then $LM = \delta x$, $ST = \delta y$ and

(area of region $LMSR$) = (area of rectangle $LMTR$) + (area of region RTS).

Now (area of the region RTS) $\approx$ (area of $\triangle RTS$) for a smoothly turning curve, i.e. (area of region RTS) $\approx \frac{1}{2}\delta x . \delta y$

$\Rightarrow$ (area of region $LMSR$) $\approx y\delta x + \frac{1}{2}\delta x . \delta y = f(x)\delta x + \frac{1}{2}\delta x . \delta y$

$\Rightarrow$ (area of region $BCQP$) = (the sum of the areas of all such strips $LMSR$ making up $BCQP$)

$$\approx \sum f(x)\delta x + \sum \tfrac{1}{2}\delta x . \delta y.$$

Obviously, the smaller the width of the strips, i.e. the smaller δx becomes or the larger n becomes, the more accurate the approximation and, in the limit as $\delta x \to 0$, we have, since $\delta y \to 0$ as $\delta x \to 0$,

(area of region $BCQP$) $= \lim_{\delta x \to 0} \sum f(x)\delta x + \lim_{\delta x \to 0} \sum \tfrac{1}{2}\delta x . \delta y.$

$= \lim_{\delta x \to 0} \sum f(x)\delta x,$ where the summations extend from $x = b$ to $x = c$.

This limit occurs frequently in mathematics. It can be shown to be equal to the definite integral $\int_b^c f(x)\,dx$. Rigorous analysis is beyond the scope of this book but, for a curve for which y is positive and increasing with x, the connection between this definite integral and the indefinite integral can be shown as follows.

Let the area of region $BCQP$ be A and the area of the elementary strip $LMSR$ be δA. Then the area of region $LMSR$ is greater than the area of rectangle $LMTR$ but less than the area of rectangle $LMSU$,

$$\Rightarrow y\delta x < \delta A < (y + \delta y)\delta x \quad \text{or} \quad y < \frac{\delta A}{\delta x} < y + \delta y.$$

In the limit as the width δx of the strip is decreased to zero, δy and δA also tend to zero. Thus

$$y + \delta y \rightarrow y, \quad \frac{\delta A}{\delta x} \rightarrow \frac{\mathrm{d}A}{\mathrm{d}x} \quad \text{and hence} \quad \frac{\mathrm{d}A}{\mathrm{d}x} = y.$$

Integrating, $\qquad A = \int \frac{\mathrm{d}A}{\mathrm{d}x}\mathrm{d}x = \int y\,\mathrm{d}x,$

i.e. the area of the finite region contained between the curve $y = \mathrm{f}(x)$, the x-axis and the ordinates at $x = b, x = c$ is given by $\int_b^c y\,\mathrm{d}x = \int_b^c \mathrm{f}(x)\,\mathrm{d}x$.

(*Note:* The integration is with respect to x and therefore the limits are x limits).

In the same way it can be shown that the area contained between the curve $y = \mathrm{f}(x)$, the y-axis and the straight lines $y = k, y = l, l > k$ is given by $\int_k^l x\,\mathrm{d}y$,

the integration being with respect to y and therefore the limits are y limits. This result is illustrated in Fig. 8.2.

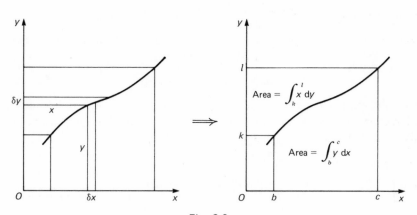

Fig. 8.2

In $\lim\limits_{\delta x \rightarrow 0} \sum\limits_{x=b}^{x=c} \mathrm{f}(x)\,\mathrm{d}x = \int_b^c \mathrm{f}(x)\,\mathrm{d}x$, the $\int$ sign may be looked upon as an elongated S (the letter used in earlier times for summation), standing for the sum of $\mathrm{f}(x)\delta x$ as $\delta x \rightarrow 0$. The limits are the range over which the summation is taken.

Example 5 Find the area in the first quadrant contained between the curve $y^2 = 4(1 - x)$ and the coordinate axes.

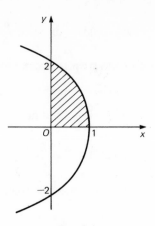

Fig. 8.3

The curve $y^2 = 4(1 - x)$ contains even powers of y only and is therefore symmetrical about the x-axis. It is, in fact, a parabola whose vertex is at $(1, 0)$ and whose axis is the x-axis, as shown in Fig. 8.3.

From above it will be seen that the area can be obtained using either an elementary strip of area $y\delta x$ or one of area $x\delta y$. Thus the area A of the shaded region is given by

$$A = \int_0^1 y\,dx \quad \text{or} \quad \int_0^2 x\,dy.$$

Using $\int_0^1 y\,dx$, $\quad A = \int_0^1 \sqrt{[4(1 - x)]}\,dx$

$$= 2\int_0^1 (1 - x)^{1/2}\,dx$$

$$= 2\left[\frac{(1 - x)^{3/2}}{(-3/2)}\right]_0^1 = -\frac{4}{3}[0 - 1] = \frac{4}{3}.$$

Using $\int_0^2 x\,dy$, $\quad A = \int_0^2 \left(1 - \frac{1}{4}y^2\right)dy$

$$= \left[y - \frac{1}{12}y^3\right]_0^2$$

$$= (2 - 8/12) - 0 = 4/3.$$

Example 6   Sketch the arc of the curve $y = \sin^2 x$ from $x = 0$ to $x = \pi$. Find the area of the region which lies beneath this arc but above the line $y = \frac{1}{4}$.

The curve can easily be sketched using the curve $y = \sin x$ and remembering that when a number between 0 and 1 is squared a smaller number is obtained. This is shown in Fig. 8.4.

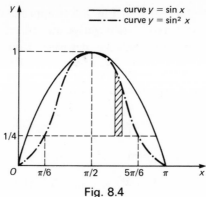

Fig. 8.4

We shall need to find where the curves $y = \sin^2 x$ and $y = \frac{1}{4}$ intersect. They do so where they have the same value of y, i.e. where $y = \frac{1}{4} = \sin^2 x$ $\Rightarrow \sin x = \pm\frac{1}{2}$. For $0 < x < \pi$, $\sin x$ can only be positive. Therefore

$$\sin x = \tfrac{1}{2} \Rightarrow x = \pi/6 \quad \text{or} \quad 5\pi/6.$$

Thus the length of an elementary strip for the area required is $\sin^2 x - \frac{1}{4}$ and the strips extend from $x = \pi/6$ to $x = 5\pi/6$. Hence, summing the areas of all the elementary strips, the required area A is given by

$$A = \int_{\pi/6}^{5\pi/6} \left(\sin^2 x - \frac{1}{4} \right) dx = \int_{\pi/6}^{5\pi/6} \left[\frac{1}{2}(1 - \cos 2x) - \frac{1}{4} \right] dx$$

$$= \left[\frac{x}{4} - \frac{1}{4} \sin 2x \right]_{\pi/6}^{5\pi/6}$$

$$= \left(\frac{5\pi}{24} - \frac{1}{4} \sin \frac{5\pi}{3} \right) - \left(\frac{\pi}{24} - \frac{1}{4} \sin \frac{\pi}{3} \right)$$

$$= \frac{4\pi}{24} + \frac{1}{4}\frac{\sqrt{3}}{2} + \frac{1}{4}\frac{\sqrt{3}}{2} = \frac{\pi}{6} + \frac{\sqrt{3}}{4}.$$

8.3 Curves leading to negative areas

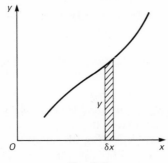

Fig. 8.5

The area of the elementary strip in Fig. 8.5 is approximately $y\delta x$. Hence, if the curve lies below the x-axis, y is negative and therefore the area derived from summing the $y\delta x$ terms will be negative. Consequently it is most important to establish the whereabouts of the curve when using the formula
$$\int_b^c y\,dx = \int_b^c f(x)\,dx$$ and if necessary take into account the sign of the area.

Example 7

(i) The area A_1 in Fig. 8.6 is given by $\qquad$ (area A_1) $= \displaystyle\int_b^c y\,dx.$

(ii) The area A_2 in Fig. 8.7 is below the x-axis. It is given by
$$\text{(area } A_2) = -\int_d^e y\,dx.$$

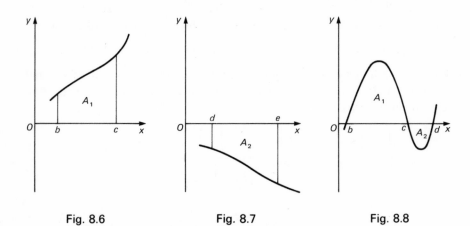

Fig. 8.6   Fig. 8.7   Fig. 8.8

(iii) The areas A_1 and A_2 in Fig. 8.8 are given by
$$A_1 = \int_b^c y\,dx, \quad A_2 = -\int_c^d y\,dx \Rightarrow A_1 + A_2 = \int_b^c y\,dx - \int_c^d y\,dx.$$

The integral, $\displaystyle\int_b^d y\,dx = \int_b^c y\,dx + \int_c^d y\,dx = A_1 - A_2$, the difference of the two areas.

A safe rule, if in doubt as to the sign to be attached, is to integrate over the separate limits arranged in order of ascending magnitude and take the numerical value of each integral.

The area of the finite region contained between the curve $y = f(x)$ shown in Fig. 8.9, the x-axis and the ordinates at $x = b$, $x = e$ is given by
$$\left|\int_b^c y\,dx\right| + \left|\int_c^d y\,dx\right| + \left|\int_d^e y\,dx\right|.$$

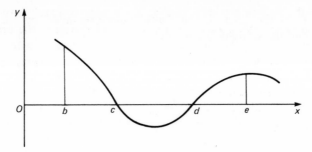

Fig. 8.9

Example 8 Show that the curve $y = \sin 3x \sin x$, $0 \leqslant x \leqslant \pi$, encloses, with the x-axis, three regions whose areas are in the ratios $1:2:1$.

The curve $y = \sin 3x \sin x$ crosses the x-axis where $y = 0 \Rightarrow \sin 3x \sin x = 0$

$$\Rightarrow \sin x = 0, \Rightarrow x = 0, \pi, \quad \text{or} \quad \sin 3x = 0 \Rightarrow x = 0, \pi/3, 2\pi/3, \pi.$$

$$\int y \, dx = \int \sin 3x \sin x \, dx$$

$$= \int \frac{1}{2}(\cos 2x - \cos 4x) \, dx$$

$$= \frac{1}{4} \sin 2x - \frac{1}{8} \sin 4x$$

$$\Rightarrow A_1 = \int_0^{\pi/3} y \, dx = \left[\frac{1}{4} \sin 2x - \frac{1}{8} \sin 4x \right]_0^{\pi/3}$$

$$= \frac{1}{4} \sin \frac{2\pi}{3} - \frac{1}{8} \sin \frac{4\pi}{3} = \frac{\sqrt{3}}{8} + \frac{\sqrt{3}}{16} = \frac{3\sqrt{3}}{16},$$

$$A_2 = \int_{\pi/3}^{2\pi/3} y \, dx = \left(\frac{1}{4} \sin \frac{4\pi}{3} - \frac{1}{8} \sin \frac{8\pi}{3} \right) - \left(\frac{1}{4} \sin \frac{2\pi}{3} - \frac{1}{8} \sin \frac{4\pi}{3} \right)$$

$$= \left(-\frac{\sqrt{3}}{8} - \frac{\sqrt{3}}{16} \right) - \frac{3\sqrt{3}}{16} = -\frac{6\sqrt{3}}{16},$$

$$A_3 = \int_{2\pi/3}^{\pi} y \, dx = \left(\frac{1}{4} \sin 2\pi - \frac{1}{8} \sin 4\pi \right) - \left(\frac{1}{4} \sin \frac{4\pi}{3} - \frac{1}{8} \sin \frac{8\pi}{3} \right)$$

$$= 0 - \left(-\frac{3\sqrt{3}}{16} \right) = \frac{3\sqrt{3}}{16},$$

$$\Rightarrow A_1 = \frac{3\sqrt{3}}{16}, \quad |A_2| = \frac{6\sqrt{3}}{16}, \quad A_3 = \frac{3\sqrt{3}}{16}$$

$$\Rightarrow A_1 : |A_2| : A_3 = 1:2:1.$$

Example 9 Find the area of the finite region bounded by the parabola $y = 2 - x^2$ and the straight line $y = x$.

In solving such a question it is essential to sketch the area required.

The equation $y = 2 - x^2$ contains even powers of x only; hence the curve is symmetrical about the y-axis.

If $x = 0$, $y = 2$; if $y = 0$, $x = \pm\sqrt{2}$. Also $2 - x^2 \leqslant 2$; hence the curve is as shown in Fig. 8.10.

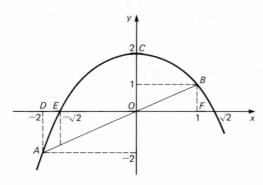

Fig. 8.10

The parabola and the straight line intersect where

$$y = 2 - x^2 = x \Rightarrow x^2 + x - 2 = 0 = (x + 2)(x - 1) \Rightarrow x = 1 \quad \text{or} \quad -2.$$

Area of region ABC = area of region $EFBC$ + area $\triangle OAD$ − area $\triangle OFB$ − area of region ADE.

$$\text{Area of region } EFBC = \int_{-\sqrt{2}}^{1} (2 - x^2)\,dx = \left[2x - \frac{1}{3}x^3 \right]_{-\sqrt{2}}^{1}$$

$$= 2 - \frac{1}{3} + 2\sqrt{2} - \frac{2\sqrt{2}}{3} = \frac{5}{3} + \frac{4\sqrt{2}}{3}.$$

$$|\text{Area of region } ADE| = -\int_{-2}^{-\sqrt{2}} (2 - x^2)\,dx = \left[2x - \frac{1}{3}x^3 \right]_{-\sqrt{2}}^{-2}$$

$$= -4 + \frac{8}{3} + 2\sqrt{2} - \frac{2\sqrt{2}}{3} = -\frac{4}{3} + \frac{4\sqrt{2}}{3}.$$

Area $\triangle OFB = \frac{1}{2} . 1 . 1 = \frac{1}{2}$.

Area $\triangle OAD = \frac{1}{2} . 2 . 2 = 2$.

$$\text{Area required} = \left(\frac{5}{3} + \frac{4\sqrt{2}}{3} \right) + 2 - \frac{1}{2} - \left(-\frac{4}{3} + \frac{4\sqrt{2}}{3} \right) = 4\frac{1}{2}.$$

Having established this result, the reader should note that it could have been obtained by considering an elementary strip of length $y_1 - y_2$ as shown in

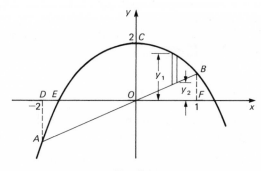

Fig. 8.11

Fig. 8.11, where $y_1 = 2 - x^2$, $y_2 = x$ and summing the areas of all such strips between $x = -2$ and $x = 1$; the signs to be attached to the areas will in fact be automatically taken into account

$$\Rightarrow \text{Area required} = \int_{-2}^{1} [(2 - x^2) - x]\,dx$$

$$= \left[2x - \frac{x^3}{3} - \frac{x^2}{2} \right]_{-2}^{1}$$

$$= 2 - \frac{1}{3} - \frac{1}{2} - \left(-4 + \frac{8}{3} - 2 \right) = 4\frac{1}{2}.$$

8.4 Areas for curves given in parametric form

When the equation of the curve is given in parametric form, i.e. $x = x(t)$, $y = y(t)$, the area is normally determined by changing the relevant integral to one in terms of the independent variable t.

Example 10 Sketch the cycloid $x = a(t - \sin t)$, $y = a(1 - \cos t)$, where a is a positive constant, for $0 \leqslant t \leqslant 2\pi$, showing that it is symmetrical about the line $x = \pi a$. Find the area of the region contained between the curve and the x-axis.

The curve $x = a(t - \sin t)$, $y = a(1 - \cos t)$ is a rather special curve. It is the curve traced out by a point on the circumference of a circle, in this case of radius a, as the circle rolls along a straight line, in this case the x-axis.

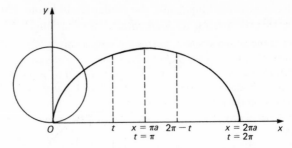

Fig. 8.12

When $t = 0$, $x = 0$, $y = 0$; thus the point on the circle under consideration is the point at the origin.

When $t = 2\pi$, $\qquad\qquad\qquad\qquad\qquad x = 2\pi a$, $\quad y = 0$.

When $t = \pi$, $\qquad\qquad\qquad\qquad\qquad\ x = \pi a$, $\quad y = 2a$.

To show that the curve is symmetrical about the line $x = \pi a$ it is necessary to show that y has the same value at points equidistant from $x = \pi a$, i.e. that $y(2\pi - t) = y(t)$.

Now $\qquad\qquad y(2\pi - t) = a[1 - \cos(2\pi - t)] = a(1 - \cos t) = y(t)$

$\Rightarrow$ curve is symmetrical about $x = \pi a$ and is as shown in Fig. 8.12.
Area of the region contained between the curve and the x-axis

$$= \int_{x=0}^{x=2\pi a} y\,dx = \int_{t=0}^{t=2\pi} y \cdot \frac{dx}{dt} \cdot dt$$

$$= \int_0^{2\pi} a(1 - \cos t) \cdot a(1 - \cos t)\,dt = a^2 \int_0^{2\pi} (1 - 2\cos t + \cos^2 t)\,dt$$

$$= a^2 \int_0^{2\pi} [1 - 2\cos t + \tfrac{1}{2}(1 + \cos 2t)]\,dt$$

$$= a^2 \left[\frac{3t}{2} - 2\sin t + \frac{1}{4}\sin 2t\right]_0^{2\pi} = 3\pi a^2.$$

Exercise 8.4

1 Find the area of the finite region bounded by the x-axis and by that part of the curve $y = \sin 2x$ between $x = 0$ and $x = \pi/2$.

2 Find the area of the finite region bounded by the curve $y^2 = 2 - x$ and the y-axis.

3 Find the area of the finite region bounded by the curve $y = x^2 - 25$ and the straight line $y = x - 13$.

4 The gradient of a curve is $6x - 3x^2$. If the curve passes through the origin find its equation and show that it cuts the x-axis again where $x = 3$. Find also the area of the finite region bounded by the curve and the x-axis. (L)

5 Calculate the area of the region enclosed between the curve $y = \cos(2x - \pi/3)$, $0 \leqslant x \leqslant 5\pi/12$, and the coordinate axes. (AEB)

6 Sketch the curve $y = \dfrac{x-2}{5+x^2}$ and show that the angle between the tangents to the curve at the points where it crosses the axes is $\tan^{-1}(2/23)$. Calculate the area of the region in the fourth quadrant bounded by the curve and the axes of coordinates. (AEB)

7 Sketch the curve $y = xe^{-x}$ for values of x from 0 to 1. Show that the curve divides the region bounded by the axes, the line $x = 1$ and the line $y = 1/e$ into two parts with areas in the ratio $(e - 2):(3 - e)$. (AEB)

8 Sketch the graph of $y^2 = x(x - 3)^2$ and find the area of the region bounded by the loop. (L)

9 Sketch the curve $y = x^3$.

A is the point $(1, 1)$ on this curve. Find the equation of the tangent to the curve at A. If this tangent meets the curve again at B, show that B is the point $(-2, 8)$.

Find the area of the finite region bounded by the arc AB of the curve and the chord AB.

8.5 Volumes; Volumes of revolution

We saw that the area under a curve could be determined by dividing the area into a series of elementary strips of width δx and summing the area of all the strips as $\delta x \to 0$.

Likewise, if we are required to determine a volume, e.g. that of a log, we can divide the volume up into a series of elementary volumes each of thickness δx. Thus, if $A(x)$ is the area of cross-section of an elementary volume of thickness δx, then the volume of the element is approximately $A(x)\delta x$. Summing all such elements as $\delta x \to 0$ we obtain the total volume V, where

$$V = \lim_{\delta x \to 0} \sum_{x=b}^{x=c} A(x)\delta x.$$

Here b and c are the beginning and end values of x for the elementary volumes.

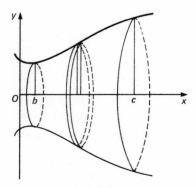

Fig. 8.13

Of particular interest is a volume of revolution as shown in Fig. 8.13, i.e. the volume obtained when the region contained between the curve $y = f(x)$, the x-axis and the ordinates at $x = b, x = c$ $(c > b)$ is revolved completely about the

x-axis. In this instance all of the areas of the cross-sections of the elementary volumes are circles, the radius of any one being that of the ordinate at the point under consideration. Thus $A(x) = \pi y^2$.

Hence the volume of revolution about the x-axis is given by

$$V_{Ox} = \lim_{\delta x \to 0} \sum_{x=b}^{x=c} \pi y^2 \delta x.$$

In exactly the same way that we showed, when considering area, the relationship between a limit of a sum of areas and the indefinite integral $\int y\,dx$, we can establish the corresponding relationship for volumes of revolution about the x-axis; namely, $\lim_{\delta x \to 0} \sum A(x)\delta x = \int_b^c \pi y^2\,dx$ where the summation extends from $x = b$ to $x = c$.

Again note the summation is along the x-axis, i.e. the integration is with respect to x and therefore the limits are x limits.

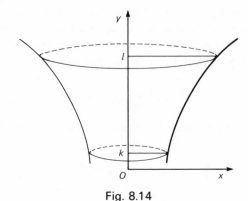

Fig. 8.14

Had the volume been a volume of revolution about the y-axis as shown in Fig. 8.14 then the elementary volume would have had a thickness δy and a circular cross-section of radius x. Hence its volume would be approximately $\pi x^2 \delta y$ and the total volume of revolution about the y-axis $V_{Oy} = \pi \int_k^l x^2\,dy$, the limits of integration being the y limits of the end points of the area under consideration.

Example 11 Find the volume generated when the region enclosed by the line $y = a$, the line $x = 3a$ and the arc of the parabola $y^2 = ax$, $a > 0$, between the points (a, a) and $(3a, \sqrt{3a})$, is rotated completely about the x-axis.

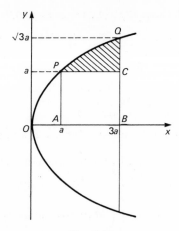

Fig. 8.15

The region under consideration is the shaded region shown in Fig. 8.15. The volume obtained when this region is revolved completely about the x-axis is the same as the volume of revolution of area *ABQP* minus the volume of revolution of area *ABCP*. Now the volume of revolution of area *ABQP* is

$$\pi \int_a^{3a} ax \, dx = \pi a \left[\frac{x^2}{2} \right]_a^{3a} = \pi a \left[\frac{9a^2}{2} - \frac{a^2}{2} \right] = 4\pi a^3.$$

Volume of revolution of area *ABCP* is the volume of a cylinder, of radius *a* and length 2a, that is, $\pi a^2 . 2a = 2\pi a^3$.

Volume of revolution of region *PCQ* is $4\pi a^3 - 2\pi a^3 = 2\pi a^3$.

Example 12   The finite region bounded by the curve $y = b(1 + e^{x/a})$, the x-axis and the ordinates at $x = 0$, $x = 2a$ is revolved completely about the x-axis. Find the volume of revolution.

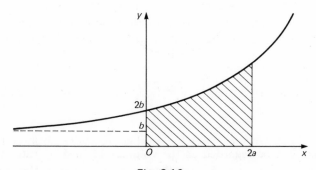

Fig. 8.16

The area to be revolved about the x-axis is as shown in Fig. 8.16.

$$\Rightarrow \text{volume generated} = \pi \int_0^{2a} y^2 \, dx$$

$$= \pi \int_0^{2a} b^2 (1 + e^{x/a})^2 \, dx$$

$$= \pi b^2 \int_0^{2a} (1 + 2e^{x/a} + e^{2x/a}) \, dx$$

$$= \pi b^2 \left[x + \frac{2e^{x/a}}{(1/a)} + \frac{e^{2x/a}}{(2/a)} \right]_0^{2a}$$

$$= \pi b^2 \left[\left(2a + 2ae^2 + \frac{a}{2}e^4 \right) - \left(0 + 2a + \frac{a}{2} \right) \right]$$

$$= \pi ab^2 \left[2e^2 + \frac{1}{2}e^4 - \frac{1}{2} \right].$$

Example 13   A hole of circular cross-section, radius r, is drilled symmetrically through a solid sphere of radius R. Find the volume removed.

The problem is essentially equivalent to finding the volume of a cap of a sphere of radius R, the height of the cap being $R - \sqrt{(R^2 - r^2)} = R - h$, say, where $h = \sqrt{(R^2 - r^2)}$. See Fig. 8.17.

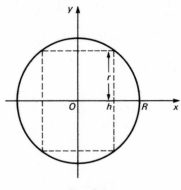

Fig. 8.17

The sphere can be regarded as the volume obtained by rotating the area within the circle $x^2 + y^2 = R^2$ completely about the x-axis. Hence,

$$\text{volume of the cap} = \int_h^R \pi y^2 \, dx$$

$$= \pi \int_h^R (R^2 - x^2) \, dx$$

$$= \pi \left[R^2 x - \frac{1}{3} x^3 \right]_h^R$$

$$= \pi \left[R^3 - \frac{1}{3} R^3 - R^2 h + \frac{1}{3} h^3 \right]$$

$$= \pi \left[\frac{2}{3} R^3 - R^2 h + \frac{1}{3} h^3 \right].$$

Thus volume removed = (volume of two such caps)
+ (volume of a cylinder radius r height $2h$)

$$= 2\pi \left[\frac{2}{3} R^3 - R^2 h + \frac{1}{3} h^3 \right] + 2\pi r^2 h, \quad \text{where } h = \sqrt{(R^2 - r^2)}.$$

8.6 The shell method

It must not be assumed that all volumes of revolution are given by the integrals $\pi \int y^2 \, dx$ or $\pi \int x^2 \, dy$. The form of the integral depends on the shape and position of the area relative to the axis about which it is to be revolved. The student is advised to build up the form of the integral by considering each time an appropriate elementary strip.

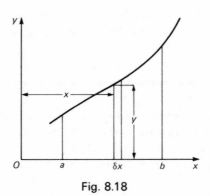

Fig. 8.18

Consider for instance the volume obtained by revolving the area contained between the curve $y = f(x)$, the x-axis and ordinates at $x = a$, $x = b$ about the y-axis. This can be obtained by the 'shell method' by considering an elementary strip which is parallel to the y-axis, is of height y and width δx, and is at a distance

x from the y-axis, as shown in Fig. 8.18. When revolved about the y-axis a cylindrical shell of height y, radius x and thickness δx is traced out, giving a volume $y \cdot 2\pi x \delta x$. The total volume of revolution is the sum of all such 'shells', i.e.

$$\text{volume} = \lim_{\delta x \to 0} \sum_{x=a}^{x=b} 2\pi xy\delta x = 2\pi \int_{x=a}^{x=b} xy\,dx.$$

It should be noted that the integration is with respect to x and therefore the limits are x limits.

Example 14 The finite region bounded by the curve $y = x^2$ and the line $y = 3$ generates various solids of revolution when rotated as follows:
(i) about the x-axis, (ii) about the line $y = -1$, (iii) about the line $x = 2$.
Form an integral to give the volume generated in each case.

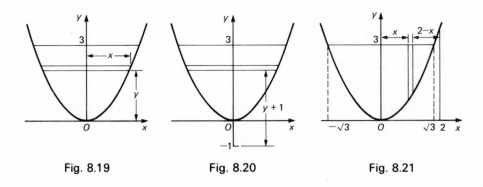

Fig. 8.19   Fig. 8.20   Fig. 8.21

(i) Consider an elementary strip of thickness δy, parallel to and distance y from the x-axis, as in Fig. 8.19. The length of the strip is $2x$, hence the volume of the elementary shell is $2\pi y\delta y \cdot 2x$

$$\Rightarrow \text{volume of revolution} = \int_{y=0}^{y=3} 4\pi xy\,dy = \int_0^3 4\pi y^{3/2}\,dy,$$

the integration being with respect to y and therefore y limits are required.
(ii) If the same elementary strip is revolved about the line $y = -1$ then the radius of the elementary cylindrical shell is $y + 1$. See Fig. 8.20. Hence the volume of the elementary shell is $2\pi(y + 1)\delta y \cdot 2x$

$$\Rightarrow \text{total volume of revolution} = \int_{y=0}^{y=3} 4\pi(y + 1)x\,dy = \int_0^3 4\pi(1 + y)y^{1/2}\,dy,$$

again the limits being y limits.
(iii) Consider an elementary strip of thickness δx, parallel to and distance x from the y-axis as shown in Fig. 8.21. The length of the strip is $3 - y$ and its distance from the axis of revolution is $2 - x$.

Hence the volume of the elementary cylindrical shell is $2\pi(2 - x)\delta x.(3 - y)$

$$\Rightarrow \text{total volume of revolution} = \int_{x=-\sqrt{3}}^{x=\sqrt{3}} 2\pi(2 - x)(3 - y)\,dx$$

$$= \int_{-\sqrt{3}}^{\sqrt{3}} 2\pi(2 - x)(3 - x^2)\,dx,$$

this time the integration being with respect to x requires x limits of $\pm\sqrt{3}$ for the elementary strips.

Exercise 8.6

1   Find the coordinates of the points of intersection of the two parabolas $y^2 = 4ax$ and $x^2 = 4ay$, $a > 0$. Show that the area of the finite region between these parabolas is $16a^2/3$.

   This region is rotated through four right angles about the x-axis. Show that the volume generated is $96\pi a^3/5$. (L)

2   Prove that the volume of a spherical cap, of height h and cut from a sphere of radius r, is $\frac{1}{3}\pi h^2(3r - h)$. (L)

   A sphere of radius r rests in an inverted, hollow, right circular cone of semi-vertical angle $30°$. Calculate in terms of r the volume contained between the sphere and the cone. (L)

3   A vase is formed by the rotation about the y-axis of the part of the curve $x = 1 + \frac{1}{2}\sin y$ between $y = 0$ and $y = 2\pi$, the base being formed by the rotation of the part of the x-axis between $x = 0$ and $x = 1$. If x and y are measured in cm, find, correct to one place of decimals, the volume of the vase. (AEB)

4   Sketch the curve $y = (x^2 - a^2)^2$, where $a > 0$. Find the area of the finite region bounded by this curve and the x-axis, and determine the volume obtained when this region is rotated through two right angles about the y-axis.

5   Sketch the curve $y = \ln(x - 2)$.

   The inner surface of a bowl is of the shape formed by rotating completely about the y-axis that part of the x-axis between $x = 0$ and $x = 3$ and that part of the curve $y = \ln(x - 2)$ between $y = 0$ and $y = 2$. The bowl is placed with its axis vertical and water is poured in. Calculate the volume of water in the bowl when it is filled to a depth $h(<2)$.

   If water is poured into the bowl at a rate of 50 cubic units per second, find the rate at which the water level is rising when the depth of the water is 1.5 units. (AEB)

6   Find the volume obtained when the finite region contained between the curve $y = \sin x$ and the x-axis from $x = 0$ to $x = \pi$ is revolved completely about (i) the x-axis, (ii) the y-axis.

7   The finite region contained in the first quadrant between the curve $y = \ln(x/a)$, the x-axis and the ordinate at $x = 2a$ is revolved completely about the y-axis to form a ring. Calculate the volume of the ring.

8.7   Mean values

We have seen that the area contained between the curve $y = f(x)$, shown in Fig. 8.22, the x-axis and the ordinates at $x = a$, $x = b$, where $b > a$, is given by

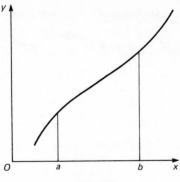

Fig. 8.22

$$\int_a^b y\,dx = \int_a^b f(x)\,dx.$$

If this area is divided by $b - a$, i.e. the range of integration, the quantity $\dfrac{1}{(b - a)}\displaystyle\int_a^b y\,dx$ gives the height of the rectangle standing on the same base, of length $(b - a)$, and having the same area as the region contained between the curve, the x-axis and the ordinates at $x = a$, $x = b$. It represents the mean height of the area and is called the mean value of y with respect to x over the range $a \leqslant x \leqslant b$.

Example 15   Given that $y = \dfrac{6x - 10}{(x - 3)^2(x + 1)}$, express y in partial fractions. Determine the mean value of y with respect to x in the range $4 \leqslant x \leqslant 6$.

$$\frac{6x - 10}{(x - 3)^2(x + 1)} = \frac{1}{x - 3} + \frac{2}{(x - 3)^2} - \frac{1}{x + 1},\ \text{by partial fractions,}$$

$$\Rightarrow \int_4^6 \frac{6x - 10}{(x - 3)^2(x + 1)}\,dx = \int_4^6 \left[\frac{1}{x - 3} + \frac{2}{(x - 3)^2} - \frac{1}{(x + 1)}\right]dx$$

$$= \left[\ln(x - 3) - \frac{2}{x - 3} - \ln(x + 1)\right]_4^6$$

$$= \ln 3 - \frac{2}{3} - \ln 7 - \ln 1 + \frac{2}{1} + \ln 5$$

$$= \ln\frac{15}{7} + \frac{4}{3}$$

$\Rightarrow$ mean value of y in the range $x = 4$ to $x = 6$ is

$$\frac{1}{6 - 4}\left[\ln\left(\frac{15}{7}\right) + \frac{4}{3}\right] = \frac{1}{2}\ln\left(\frac{15}{7}\right) + \frac{2}{3}.$$

8.8 Root mean square value

The root mean square value or R.M.S. value of $y = f(x)$ over a given range is defined as the square root of the mean value of y^2 over that range. R.M.S. of y over the range $a \leqslant x \leqslant b$ is given by

$$\sqrt{\left[\frac{1}{(b-a)} \int_a^b y^2 \, dx\right]} = \sqrt{\left[\frac{1}{(b-a)} \int_a^b [f(x)]^2 \, dx\right]}.$$

Example 16 Find the root mean square value of the current $i = I \sin \omega t$ over the range $t = 0$ to $t = \pi/\omega$.

Mean value of i^2 is $\quad \dfrac{1}{(\pi/\omega)} \displaystyle\int_0^{\pi/\omega} I^2 \sin^2 \omega t \, dt$

$$= \frac{\omega}{2\pi} \int_0^{\pi/\omega} I^2 (1 - \cos 2\omega t) \, dt$$

$$= \frac{\omega I^2}{2\pi} \left[t - \frac{1}{2\omega} \sin 2\omega t \right]_0^{\pi/\omega} = \frac{I^2}{2}$$

$$\Rightarrow \text{R.M.S. of } i \text{ is } \frac{I}{\sqrt{2}}.$$

8.9 First moment of area

Consider a region of area A and a small element, of area δA_p, of this region containing the point $P(x_p, y_p)$, as shown in Fig. 8.23.

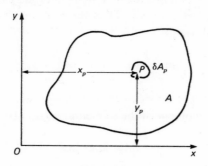

Fig. 8.23

Then the first moment of the element of area δA_p about the y-axis is defined as the product of the area δA_p and its perpendicular distance from the y-axis, i.e. as $x_p \, \delta A_p$.

The first moment of the total area of the region about the y-axis is thus $\displaystyle \lim_{\delta A_p \to 0} \sum_A x_p \delta A_p$.

In evaluating this limit it is convenient to choose a suitable form for the elementary area δA_p. Thus for a region contained between the curve $y = f(x)$, the x-axis and the ordinates at $x = a$, $x = b$, where $b > a$, it is appropriate to choose δA_p in the form of an elementary strip of width δx and parallel to the y-axis. All points on the strip are then at an equal distance x from the y-axis.

Hence the first moment of the area shown in Fig. 8.24 about the y-axis is given by $\displaystyle\lim_{\delta x \to 0} \sum_{x=a}^{x=b} x \cdot y \delta x = \int_a^b xy\,dx.$

If the first moment of the area about the x-axis is required, then it is necessary to calculate initially the first moment of the elementary strip about the x-axis. This can be done as follows.

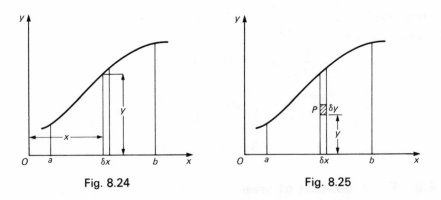

Fig. 8.24   Fig. 8.25

Consider Fig. 8.25 which shows an elementary rectangular area $\delta x \cdot \delta y$ at P distance y from the x-axis. Then the first moment of this elementary rectangular area about the x-axis is $y \cdot \delta x \cdot \delta y$. Thus the first moment of the elementary strip shown about the x-axis is

$$\lim_{\delta y \to 0} \sum_{y=0}^{y=y} y \delta x \cdot \delta y = \delta x \int_0^y y\,dy,$$

where the δx can be taken outside the integral sign since, along the strip, δx is constant, i.e. the first moment of the elementary strip about the x-axis is

$$\delta x \cdot \frac{y^2}{2} = \tfrac{1}{2} y^2 \delta x,$$

$\Rightarrow$ the first moment of the area about the x-axis is given by

$$\lim_{\delta x \to 0} \sum_{x=a}^{x=b} \tfrac{1}{2} y^2 \delta x = \int_a^b \tfrac{1}{2} y^2\,dx.$$

Example 17 Calculate the first moment of area of the finite region enclosed between the curve $y = a \cos (2x - \pi/3)$, $0 \leqslant x \leqslant 5\pi/12$, $a > 0$, and the coordinate axes about (i) the y-axis, (ii) the x-axis.

The curve $y = a \cos (2x - \pi/3)$, $0 \leqslant x \leqslant 5\pi/12$ is shown in Fig. 8.26.

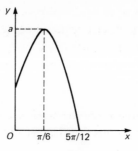

Fig. 8.26

First moment about the y-axis $= \displaystyle\int_0^{5\pi/12} xy\,\mathrm{d}x$

$$= \int_0^{5\pi/12} ax\cos(2x - \pi/3)\,\mathrm{d}x$$

$$= a\left[x\frac{\sin(2x - \pi/3)}{2}\right]_0^{5\pi/12}$$

$$\quad - a\int_0^{5\pi/12} \tfrac{1}{2}\sin(2x - \pi/3).1\,\mathrm{d}x$$

$$= \frac{5\pi a}{24}\sin\frac{\pi}{2} + a\left[\frac{1}{4}\cos\left(2x - \frac{\pi}{3}\right)\right]_0^{5\pi/12}$$

$$= \frac{5\pi a}{24} + \frac{a}{4}\cos\frac{\pi}{2} - \frac{a}{8}$$

$$= \frac{5\pi a}{24} - \frac{a}{8}.$$

First moment about the x-axis $= \displaystyle\int_0^{5\pi/12} \tfrac{1}{2}y^2\,\mathrm{d}x$

$$= \tfrac{1}{2}\int_0^{5\pi/12} a^2\cos^2(2x - \pi/3)\,\mathrm{d}x$$

$$= \frac{a^2}{4}\int_0^{5\pi/12}\left[\cos\left(4x - \frac{2\pi}{3}\right) + 1\right]\mathrm{d}x$$

$$= a^2\left[\frac{1}{16}\sin\left(4x - \frac{2\pi}{3}\right) + \frac{x}{4}\right]_0^{5\pi/12}$$

$$= a^2\left[\frac{1}{16}\sin\pi - \frac{1}{16}\sin\left(-\frac{2\pi}{3}\right) + \frac{5\pi}{48} - 0\right]$$

$$= a^2\left(\frac{1}{16}\cdot\frac{\sqrt{3}}{2} + \frac{5\pi}{48}\right) = \left(\frac{\sqrt{3}}{32} + \frac{5\pi}{48}\right)a^2.$$

8.10 Centroids

If the first moment about the y-axis of the region contained between the curve $y = f(x)$, the x-axis and the ordinates at $x = a$, $x = b$ is equated to the area, A, of the region times a constant, say $\bar{X}$, then $\bar{X}$ is said to be the x-coordinate of the *centroid* of the area.

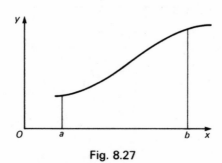

Fig. 8.27

For the curve shown in Fig. 8.27 this gives

$$\bar{X}.\int_a^b y\,dx = \int_a^b x.y\,dx \Rightarrow \bar{X} = \int_a^b xy\,dx \Big/ \int_a^b y\,dx.$$

Similarly, if the first moment of the region about the x-axis is equated to the area, A, of the region times a constant $\bar{Y}$, then $\bar{Y}$ is said to be the y-coordinate of the *centroid* of the region

$$\Rightarrow \bar{Y}\int_a^b y\,dx = \int_a^b \tfrac{1}{2}y^2\,dx \Rightarrow \bar{Y} = \int_a^b \tfrac{1}{2}y^2\,dx \Big/ \int_a^b y\,dx.$$

For a lamina of uniform thickness, having the same shape as the region of area A, then $(\bar{X}, \bar{Y})$ are the coordinates of the centre of gravity of the lamina. If the lamina does not have uniform thickness or density then it will be necessary to modify the form of the integral in order to take this into account. As with volumes, the student should learn to 'build up' his own formula by considering an appropriate elementary area.

Example 18   Sketch the curve $y^2 = \dfrac{ax^2}{a - x}$ from $x = 0$ to $x = a$, where a is positive. Show that the area of the region between the curve and the line $x = a$ is $8a^2/3$ and that the centroid of this region is distant $a/5$ from the line $x = a$.

The equation of the curve contains even powers of y only $\Rightarrow$ curve is symmetrical about x-axis.

$$x = 0, y = 0; \quad x \to a \Rightarrow y \to \pm\infty.$$

$$2y\frac{dy}{dx} = \frac{(a - x)2ax - ax^2(-1)}{(a - x)^2}$$

$$\Rightarrow \frac{dy}{dx} = \frac{ax(2a - x)}{(a - x)^2} \cdot \frac{1}{2} \frac{\sqrt{(a - x)}}{x\sqrt{a}} = \frac{(2a - x)\sqrt{a}}{2(a - x)^{3/2}}$$

$$\Rightarrow \frac{dy}{dx} = \frac{2a\sqrt{a}}{2a^{3/2}} = 1 \quad \text{when } x = 0.$$

The curve is shown in Fig. 8.28.

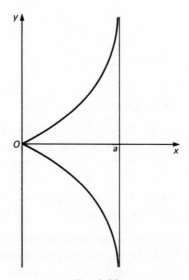

Fig. 8.28

The area A of the region between the curve and the line $x = a$ is

$$2\int_0^a y\,dx = 2\int_0^a \frac{x\sqrt{a}}{\sqrt{(a - x)}}\,dx.$$

Let $x = a \sin^2 \theta \Rightarrow dx = 2a \sin \theta \cos \theta \, d\theta$

$$\Rightarrow A = 2\int_0^{\pi/2} \frac{\sqrt{a} \cdot a \sin^2 \theta \cdot 2a \sin \theta \cos \theta}{\sqrt{(a - a \sin^2 \theta)}}\,d\theta$$

$$= 4a^2 \int_0^{\pi/2} \sin^3 \theta \, d\theta$$

$$= -4a^2 \int_0^{\pi/2} (1 - \cos^2 \theta)d(\cos \theta)$$

$$= -4a^2 \left[\cos \theta - \frac{1}{3}\cos^3 \theta\right]_0^{\pi/2} = 4a^2 \cdot \frac{2}{3} = \frac{8a^2}{3}.$$

First moment of the area about y-axis is $I = 2 \displaystyle\int_0^a xy\,dx = 2 \int_0^a \dfrac{x^2\sqrt{a}}{\sqrt{(a-x)}}\,dx$.

$$x = a\sin^2\theta \Rightarrow I = 4\int_0^{\pi/2} a^3\sin^5\theta\,d\theta$$

$$= -4a^3\int_0^{\pi/2}(1-\cos^2\theta)^2\,d(\cos\theta)$$

$$= -4a^3\left[\cos\theta - \frac{2}{3}\cos^3\theta + \frac{1}{5}\cos^5\theta\right]_0^{\pi/2}$$

$$= 4a^3\left(1 - \frac{2}{3} + \frac{1}{5}\right) = \frac{32a^3}{15}.$$

Hence
$$\frac{8a^2}{3}\bar{X} = \frac{32a^3}{15} \Rightarrow \bar{X} = \frac{4a}{5}$$

$\Rightarrow$ distance of centroid from line $x = a$ is $a - 4a/5 = a/5$.

The same procedure can also be used to find the centres of gravity of solids of revolution except that first moments must be taken about a plane rather than an axis.

Example 19 Find the position of the centre of gravity of a right circular cone of uniform density ρ, height h and base radius r.

By symmetry the centre of gravity of the cone must lie on the axis of the cone. Choose the axes as shown in Fig. 8.29 so that the vertex of the cone is at the origin and the axis of the cone coincides with the x-axis.

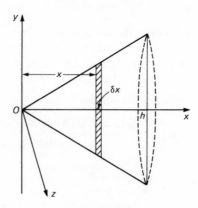

Fig. 8.29

Consider an elementary disc as shown, radius y, thickness δx and centre distant x from the origin.

The centre of gravity of this disc lies on the x-axis. Hence the first moment of the disc about the yz plane $= x \cdot \rho \cdot \pi y^2 \delta x$

$\Rightarrow$ total moment of the cone about the yz plane is $M = \displaystyle\int_0^h \rho \pi x y^2 \, dx$.

But, from similar triangles, $y/x = r/h \Rightarrow y = rx/h$

$$\Rightarrow M = \int_0^h \rho \pi r^2 \frac{x^3}{h^2} dx = \rho \frac{\pi r^2}{h^2} \left[\frac{x^4}{4} \right]_0^h = \tfrac{1}{4} \rho \pi r^2 h^2.$$

But $\quad M = \bar{X} \cdot \text{total mass of the cone} = \bar{X} \cdot \tfrac{1}{3} \pi \rho r^2 h$

$$\Rightarrow \bar{X} = (\tfrac{1}{4} \rho \pi r^2 h^2)/(\tfrac{1}{3} \rho \pi r^2 h) = 3h/4.$$

Thus the centre of gravity of the cone lies on the axis at a distance $3h/4$ from the vertex.

Exercise 8.10

1   The mean value of $\sin \theta + k \sin 2\theta$ over the range $\theta = 0$ to $\theta = \pi/2$ is M. The root mean square value over the same range is N. If $3N^2 = 4M$, determine the possible values of k. (AEB)

2   Sketch the curve $y = \sin 3x$ for $0 \leqslant x \leqslant \pi$ and calculate the ratio of the mean value to the root mean square value of y over this range.

3   Sketch the curve $y = \dfrac{1 + x}{2x + x^2}$ and find

   (i) the area of the finite region contained between the curve, the x-axis and the ordinates $x = 1$ and $x = 2$,

   (ii) the distance of the centroid of this region from the y-axis.

4   A lens has the shape formed by revolving the region contained between the parabolas $y^2 = 4x$ and $y^2 = 4(3 - 2x)$ about the x-axis through two right angles. Find the volume of the lens and determine the distance of the centre of gravity of the lens from the y-axis. (AEB)

5   Sketch the curve $y^2 = a(a - x)$, where $a > 0$.

   Find the equation of the tangent and of the normal to the curve at the point $(0, a)$.

   Find also the x-coordinate of the centroid of the region in the first quadrant bounded by this tangent, the curve and the x-axis. (L)

6   Sketch the curve $y = \dfrac{2 + x}{1 - x^2}$ and find the area of the region contained between the curve and the straight line $y = 2$. Determine the distance of the centroid of this region from the y-axis.

7   Sketch the cycloid $x = t - \sin t$, $y = 1 - \cos t$ for $0 \leqslant t \leqslant 2\pi$, showing that it is symmetrical about the line $x = \pi$. Find the centroid of the region contained between this curve and the x-axis.

8   Sketch the curve $y^2 = \dfrac{4}{x^2 + 9}$ and find the area of the region bounded by the two branches of the curve, the y-axis and the line $x = 3$. Find the first moment of this region about the y-axis. (AEB)

9 Properties of definite integrals

9.1 Elementary properties

The following properties of definite integrals should be noted.

(1) The value of a definite integral does not depend upon the letter chosen to denote the variable. It depends only on the form of the integrand and the limits. For example,

$$\int_0^1 x^2 \, dx = \int_0^1 y^2 \, dy = \int_0^1 t^2 \, dt.$$

(2) $\displaystyle\int_a^b f(x) \, dx = -\int_b^a f(x) \, dx.$

i.e. interchanging the limits changes the sign of the definite integral.

This can be shown as follows. Let the indefinite integral of $f(x)$ be $F(x)$, i.e. $\displaystyle\int f(x) \, dx = F(x)$. Then

$$\int_a^b f(x) \, dx = \left[F(x) \right]_a^b = F(b) - F(a),$$

$$\int_b^a f(x) \, dx = \left[F(x) \right]_b^a = F(a) - F(b) = - \left[F(b) - F(a) \right] = - \int_a^b f(x) \, dx.$$

(3) $\displaystyle\int_a^b f(x) \, dx = \int_a^c f(x) \, dx + \int_c^b f(x) \, dx = \int_a^c f(x) \, dx + \int_c^d f(x) \, dx + \int_d^b f(x) dx,$

i.e. a definite integral can be divided up into two or more integrals of the same function.

This can be shown as follows.

$$\text{RHS} = \int_a^c f(x) \, dx + \int_c^d f(x) \, dx + \int_d^b f(x) \, dx$$

$$= \left[F(x) \right]_a^c + \left[F(x) \right]_c^d + \left[F(x) \right]_d^b$$

$$= F(c) - F(a) + F(d) - F(c) + F(b) - F(d)$$

$$= F(b) - F(a)$$

$$= \int_a^b f(x) \, dx.$$

(4) If $f(x)$ is an even function, i.e. $f(-x) = f(x)$, then

$$\int_{-a}^{a} f(x)\,dx = 2\int_{0}^{a} f(x)\,dx.$$

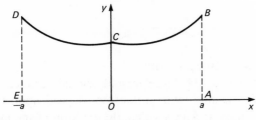

Fig. 9.1

Consider the graph of $y = f(x)$. If $f(x) = f(-x)$, then the curve is symmetrical about the y-axis. Let its graph be as shown in Fig. 9.1. Then

$$\int_{0}^{a} f(x)\,dx = \text{area of region } OABC,$$

$$\int_{-a}^{0} f(x)\,dx = \text{area of region } OCDE,$$

and, because of the symmetry about the y-axis, the areas of these regions are equal,

i.e. $$\int_{0}^{a} f(x)\,dx = \int_{-a}^{0} f(x)\,dx.$$

Hence $$\int_{-a}^{a} f(x)\,dx = \int_{-a}^{0} f(x)\,dx + \int_{0}^{a} f(x)\,dx, \quad \text{by property (3)}$$

$$= 2\int_{0}^{a} f(x)\,dx.$$

(5) If $f(x)$ is an odd function, i.e. $f(-x) = -f(x)$, then

$$\int_{-a}^{a} f(x)\,dx = 0.$$

If $f(-x) = -f(x)$, the graph of $y = f(x)$ is symmetrical about the *origin*. Let it be of the form shown in Fig. 9.2. Then

$$\int_{0}^{a} f(x)\,dx = \text{area of region } OABC,$$

$$\int_{-a}^{0} f(x)\,dx = -\text{area of region } ODEF, \text{ negative since the}$$
$$\text{area is below the } x\text{-axis.}$$

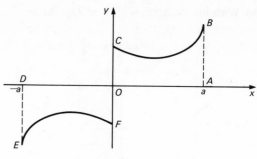

Fig. 9.2

Because of symmetry about the origin the areas of these regions are equal,

i.e.
$$\int_0^a f(x)\,dx = -\int_{-a}^0 f(x)\,dx.$$

Hence
$$\int_{-a}^a f(x)\,dx = \int_{-a}^0 f(x)\,dx + \int_0^a f(x)\,dx = 0.$$

Example 1 Given that $\int_0^{\pi/2} \sin^4 x\,dx = 3\pi/16$, evaluate $\int_0^{\pi} \sin^4 x\,dx$.

By property (3),
$$\int_0^{\pi} \sin^4 x\,dx = \int_0^{\pi/2} \sin^4 x\,dx + \int_{\pi/2}^{\pi} \sin^4 x\,dx.$$

Now for $\int_{\pi/2}^{\pi} \sin^4 x\,dx$, let $x = \pi - t$. Then

$$\int_{\pi/2}^{\pi} \sin^4 x\,dx = \int_{\pi/2}^0 \sin^4(\pi - t)(-1)\,dt$$

$$= \int_0^{\pi/2} \sin^4(\pi - t)\,dt \quad \text{by property (2),}$$

$$= \int_0^{\pi/2} \sin^4 t\,dt = \int_0^{\pi/2} \sin^4 x\,dx, \quad \text{by property (1),}$$

$$\Rightarrow \int_0^{\pi} \sin^4 x\,dx = 2\int_0^{\pi/2} \sin^4 x\,dx = 2.3\pi/16 = 3\pi/8.$$

Example 2 Evaluate $\int_{-2}^2 x^3\sqrt{(4 - x^2)}\,dx$.

$f(x) = x^3\sqrt{(4 - x^2)}$ is an odd function, since

$$f(-x) = (-x)^3\sqrt{[4 - (-x)^2]} = -x^3\sqrt{(4 - x^2)} = -f(x),$$

$$\Rightarrow \int_{-2}^2 x^3\sqrt{(4 - x^2)}\,dx = 0.$$

Example 3 Without attempting to evaluate the integrals, determine whether the following integrals are positive, negative or zero, giving reasons for your answers:

(i) $\displaystyle\int_0^\pi \sin^3 x \cos^2 x \, dx,$ (ii) $\displaystyle\int_{-1}^1 \frac{e^x + e^{-x}}{e^x - e^{-x}} dx,$ (iii) $\displaystyle\int_{-2}^2 |x| e^{-3|x|} dx.$

(i) $\displaystyle\int_0^\pi \sin^3 x \cos^2 x \, dx.$

For $0 < x < \pi$, $\sin^3 x \cos^2 x > 0 \Rightarrow$ since the integrand is positive, the integral must be positive.

(ii) $\displaystyle\int_{-1}^1 \frac{e^x + e^{-x}}{e^x - e^{-x}} dx.$

$$f(x) = \frac{e^x + e^{-x}}{e^x - e^{-x}} \Rightarrow f(-x) = \frac{e^{-x} + e^x}{e^{-x} - e^x} = -\frac{e^x + e^{-x}}{e^x - e^{-x}} = -f(x)$$

$\Rightarrow$ the integral is zero.

(iii) $\displaystyle\int_{-2}^2 |x| e^{-3|x|} dx.$

For $-2 \leqslant x \leqslant 2$, $|x| \geqslant 0$ and $e^{-3|x|} > 0$
$\Rightarrow$ for $-2 \leqslant x \leqslant 2$, $|x| e^{-3|x|} \geqslant 0$.
$\Rightarrow$ since the integrand is positive except when $x = 0$, the integral must be positive.

Exercise 9.1

1 Evaluate (a) $\displaystyle\int_{-1}^1 \frac{x^5 \, dx}{1 + x^2}$, (b) $\displaystyle\int_{-\pi/4}^{\pi/4} \sec^2 x \, dx$, (c) $\displaystyle\int_{-\pi/2}^{\pi/2} \sin^4 x \cos^3 x \, dx$,

(d) $\displaystyle\int_{-\pi}^{\pi} \frac{\sin^3 x}{1 + \cos^2 x} dx.$

2 Use the substitution $x = t + a$ to show that $\displaystyle\int_a^b f(x) \, dx = \int_0^{b-a} f(x + a) \, dx.$

Hence evaluate $\displaystyle\int_{\pi/2}^\pi \sin^3 x \, dx.$

3 Use the substitution $x = 2t$ to show that

$$\int_0^\pi f(x) \, dx = 2 \int_0^{\pi/2} f(2x) \, dx.$$

Hence evaluate $\displaystyle\int_0^\pi \sin^2 \left(\frac{x}{2}\right) \cos^3 \left(\frac{x}{2}\right) dx.$

4 Let $I = \displaystyle\int_0^\pi \frac{x}{1 + \sin x}\,dx$. Show, by means of the substitution $y = \pi - x$, that

$I = \pi \displaystyle\int_0^{\pi/2} \frac{1}{1 + \sin x}\,dx$. Hence, by means of the substitution $t = \tan(\tfrac{1}{2}x)$, or otherwise, evaluate I.

(L)

5 Without attempting to evaluate the integrals, determine whether the following integrals are positive, negative or zero, giving reasons for your answers:

(i) $\displaystyle\int_0^\pi x^2 \cos x\,dx$,   (ii) $\displaystyle\int_0^\pi \sin^2 x \cos^3 x\,dx$,   (iii) $\displaystyle\int_{1/2}^1 e^{-x}\ln x\,dx$.

(L)

6 The function f is such that $f(x + \pi) = f(x)$ for all values of x. In the interval $0 \leqslant x < \pi$, $f(x) = x - \sin x$. Sketch the curve $y = f(x)$ for $-2\pi \leqslant x \leqslant 2\pi$, and state all the values of x for which the function f is discontinuous. Evaluate the integrals (a) $\displaystyle\int_{-\pi/2}^{\pi/2} f(x)\,dx$,

(b) $\displaystyle\int_0^{3\pi/2} f(x)\,dx$.

(L)

7 Evaluate

(a) $\displaystyle\int_0^1 \frac{1}{(2x + 1)^2}\,dx$,   (b) $\displaystyle\int_0^1 xe^{2x}\,dx$,   (c) $\displaystyle\int_0^1 \frac{2 - x}{\sqrt{(4 - x^2)}}\,dx$.

Explain, very briefly, why, in all three cases, you would expect your answers to be positive.

(L)

10 Approximate integration

10.1 Introduction

It is not always possible when given the integral $\int f(x)\,dx$ to determine, either by substitution or otherwise, the function $F(x)$ whose derivative is $f(x)$. Consequently, we are reluctantly forced to accept that we cannot carry out the integration. In fact, there are many more functions that we cannot integrate than there are functions that we can integrate.

However, when the integral has limits, we can make use of the geometrical significance of the integral (i.e. that its value is the area of the region contained between the curve $y = f(x)$, the x-axis and the ordinates at the values of x given by the limits) to determine its value approximately.

At worst, we can resort to plotting the curve $y = f(x)$ on graph paper and then counting the relevant squares. Such a method is, however, rather tedious and unless large scales and careful plotting are undertaken it is not very accurate. Consequently, it is better to use either the trapezoidal rule or Simpson's Rule.

10.2 The trapezoidal rule

Consider the integral

$$\int_a^b f(x)\,dx.$$

For the trapezoidal rule the range of the integral, i.e. $b - a$, is divided into, say, n equal parts, each of length $h = \dfrac{b - a}{n}$. If ordinates of length $y_1, y_2, y_3, \ldots, y_{n+1}$ are then erected as shown in Fig. 10.1 at $x = a$, $x = a + h$,

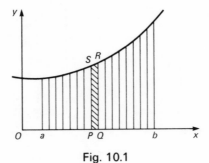

Fig. 10.1

$x = a + 2h, \ldots, x = a + nh = b$, respectively, the area contained between the curve $y = f(x)$, the x-axis and ordinates at $x = a$, $x = b$ can then be regarded as made up of the n trapeziums such as $PQRS$. The smaller the interval h, i.e. the more constituent trapeziums there are, the closer the approximation.

Since the area of a trapezium is half the sum of the parallel sides times the perpendicular distance between them, this gives

$$\int_a^b f(x)\,dx \approx \tfrac{1}{2}(y_1 + y_2)h + \tfrac{1}{2}(y_2 + y_3)h + \tfrac{1}{2}(y_3 + y_4)h + \ldots + \tfrac{1}{2}(y_n + y_{n+1})h$$

$$= h[\tfrac{1}{2}(y_1 + y_{n+1}) + (y_2 + y_3 + y_4 + \ldots + y_{n-1} + y_n)].$$

In words,

$$\int_a^b f(x)\,dx = \text{(interval width)} \times \text{(half the sum of the first and last ordinates}$$
$$+ \text{ sum of the remaining ordinates)}.$$

In practice, of course, it is not necessary to construct the trapeziums – it is only necessary to calculate the interval width and the length of the ordinates.

Example 1   Evaluate, using the trapezoidal rule, $\displaystyle\int_0^{\pi/2} \sqrt{(\sin x)}\,dx$.

Let the range of $\pi/2$ be divided into 9 intervals each equal to the radian measure of $10°$. The work is easily carried out when a calculator is used.

$y_1 = \sqrt{(\sin 0°)} \quad = 0\cdot0000$
$y_2 = \sqrt{(\sin 10°)} \qquad = 0\cdot4167$
$y_3 = \sqrt{(\sin 20°)} \qquad = 0\cdot5848$
$y_4 = \sqrt{(\sin 30°)} \qquad = 0\cdot7071$
$y_5 = \sqrt{(\sin 40°)} \qquad = 0\cdot8017$
$y_6 = \sqrt{(\sin 50°)} \qquad = 0\cdot8752$
$y_7 = \sqrt{(\sin 60°)} \qquad = 0\cdot9306$
$y_8 = \sqrt{(\sin 70°)} \qquad = 0\cdot9694$
$y_9 = \sqrt{(\sin 80°)} \qquad = 0\cdot9924$
$y_{10} = \sqrt{(\sin 90°)} = 1\cdot0000$
$\qquad\qquad\qquad\quad \overline{1\cdot0000 \quad 6\cdot2779}$

$\Rightarrow \displaystyle\int_0^{\pi/2} \sqrt{(\sin x)}\,dx$

$= \dfrac{\pi}{18}[0\cdot5000 + 6\cdot2779]$

$= \dfrac{\pi}{18} \times 6\cdot7779$

$\approx 1\cdot183.$

Since our numerical method is not exact, we give the result to 3 significant figures, as $1\cdot18$.

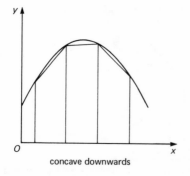

concave downwards

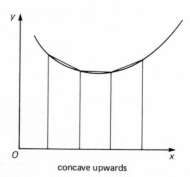

concave upwards

Fig. 10.2

64   *Integration*

As will be seen from graphical considerations, Fig. 10.2, if the curve of the integrand is concave downwards the trapezoidal rule estimate will be less than the actual value, whilst if the curve is concave upwards the trapezoidal rule estimate will be greater than the actual value. A more accurate approximation is usually obtained when the arc of the given curve is replaced by a parabolic arc rather than by the straight line sections. The formula of this better method is known as 'Simpson's rule'.

10.3 Simpson's rule

Consider three points A, B and C on the curve of the integrand as shown in Fig. 10.3, such that $A \equiv (-h, y_1), B \equiv (0, y_2), C \equiv (h, y_3)$. Through these points can also be drawn an arc of the parabola $y = ax^2 + bx + c$, since the constants a, b and c can be determined from the fact that the coordinates of A, B and C satisfy the equations

$$y_1 = ah^2 - bh + c, \quad y_2 = c, \quad y_3 = ah^2 + bh + c.$$

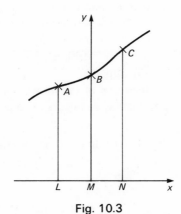

Fig. 10.3

The area of the region contained between the parabola, the x-axis and the ordinates through A and C is given by

$$\int_{-h}^{h} (ax^2 + bx + c)\,dx = \left[\frac{ax^3}{3} + \frac{bx^2}{2} + cx \right]_{-h}^{h}$$

$$= \left[\frac{ah^3}{3} + \frac{bh^2}{2} + ch \right] - \left[-\frac{ah^3}{3} + \frac{bh^2}{2} - ch \right]$$

$$= \tfrac{1}{3}h[2ah^2 + 6c]$$

$$= \tfrac{1}{3}h[(2ah^2 + 2c) + 4c]$$

$$= \tfrac{1}{3}h[y_1 + y_3 + 4y_2].$$

The fact that B lies on the y-axis makes no difference to the form of the answer. Through any three points A, B, C, where $A \equiv (x, y_1)$, $B \equiv (x + h, y_2)$,

$C \equiv (x + 2h, y_3)$, can be drawn a parabola such that the area of the region contained between the parabola, the x-axis and the ordinates through A and C is given by $\frac{1}{3}h(y_1 + 4y_2 + y_3)$. This is the simplest form of Simpson's Rule.

Hence, for a general application of Simpson's rule, the region whose area is required is divided into an even number of strips, say $2n$, by $(2n + 1)$ equidistant ordinates of lengths $y_1, y_2, y_3, \ldots y_{2n+1}$. Then, if h is the width of each of the strips, the required area can be estimated by summing the areas of the adjacent pairs of strips.

$$
\begin{aligned}
\int_a^b f(x)\,dx &= \tfrac{1}{3}h(y_1 + 4y_2 + y_3) + \tfrac{1}{3}h(y_3 + 4y_4 + y_5) + \ldots \\
&\quad + \tfrac{1}{3}h(y_{2n-1} + 4y_{2n} + y_{2n+1}) \\
&= \tfrac{1}{3}h[(y_1 + y_{2n+1}) + 4(y_2 + y_4 + \ldots + y_{2n}) \\
&\quad + 2(y_3 + y_5 + \ldots + y_{2n-1})],
\end{aligned}
$$

i.e. (one third the width interval) × (sum of the first and last ordinates
+ 4 times the sum of all the even ordinates
+ 2 times the sum of all the remaining odd ordinates).

This is Simpson's rule as usually applied.

Students are warned that the form of the above statement is for when the first ordinate is labelled y_1 and *not* y_0 as occurs in some texts.

Example 2   Use Simpson's rule with five ordinates to obtain an approximate value of $\int_1^2 x \lg x\,dx$. $[\lg x = \log_{10} x]$

The range of integration is $2 - 1 = 1$. Five ordinates gives four strips ⇒ strip width $= \frac{1}{4}$.

$$
\begin{array}{llll}
x_1 = 1\cdot00 & y_1 = 1\cdot00 \lg 1\cdot00 = 0 & & \\
x_2 = 1\cdot25 & y_2 = 1\cdot25 \lg 1\cdot25 & = 0\cdot1211 & \\
x_3 = 1\cdot50 & y_3 = 1\cdot50 \lg 1\cdot50 & & = 0\cdot2641 \\
x_4 = 1\cdot75 & y_4 = 1\cdot75 \lg 1\cdot75 & = 0\cdot4253 & \\
x_5 = 2\cdot00 & y_5 = 2\cdot00 \lg 2\cdot00 = 0\cdot6021 & & \\
\end{array}
$$

$$= 0\cdot6021 \qquad 0\cdot5464 \qquad 0\cdot2641$$

$$
\begin{aligned}
\Rightarrow \int_1^2 x \lg_{10} x\,dx &= \tfrac{1}{3}\cdot\tfrac{1}{4}[0\cdot6021 + 4 \times 0\cdot5464 + 2 \times 0\cdot2641] \\
&= \tfrac{1}{12}[0\cdot6021 + 2\cdot1856 + 0\cdot5282] \\
&= 0\cdot2763 \\
&= 0\cdot276 \text{ to 3 significant figures.}
\end{aligned}
$$

Exercise 10.3

1. Use the trapezoidal rule with five ordinates to find an approximate value for $\int_0^{1/2} \frac{1}{\sqrt{(1-x^2)}} dx$.

    Hence, find an approximate value for π.

2. Use the trapezoidal rule with ten intervals to evaluate approximately $\int_0^1 \sqrt{(1+x^2)} dx$.

    Work to four decimal places.

3. Apply Simpson's rule using five ordinates to find an approximate value of $\int_0^\pi \sin^{3/2} x \, dx$.

4. Sketch the curve $y = (x-2)(x-6)^2$ and hence sketch the curve $y^2 = (x-2)(x-6)^2$. Use Simpson's rule with five ordinates to obtain an approximate value of the area of the region in the first quadrant enclosed between the curve $y^2 = (x-2)(x-6)^2$, the x-axis and the lines $x = 3$, $x = 5$. What is the mean value of the ordinate y in the first quadrant in the range $3 \leqslant x \leqslant 5$?

5. A curve passes through the points whose coordinates are given in the table below.

x	0	1	2	3	4	5	6
y	0	1·8	3·6	4·1	3·5	2·6	0

    Find, using Simpson's rule with an interval of 1·0, (i) the area of the region contained between the curve and the x-axis, (ii) the coordinates of the centroid of this region.

6. Tabulate values of $f(x) = \sqrt{[27 + (x-3)^2]}$ for integral values of x from 0 to 6 inclusive and sketch the graph of $y = f(x)$ for the interval $0 \leqslant x \leqslant 6$.

    Given that $F(t) \equiv \int_0^t f(x) dx$, use Simpson's rule and the calculated values of $f(x)$ to estimate $F(2)$ and $F(6)$.

7. Evaluate $\int_0^{0·4} x \ln(1+x) dx$ using Simpson's rule with five ordinates and correcting your answer to two significant figures. Show clearly how your answer has been obtained. (L)

8. Show that the formula
    $$\int_0^{2h} f(x) dx \approx \frac{h}{3}[f(0) + 4f(h) + f(2h)]$$
    is exact when $f(x)$ is a cubic polynomial in x. Use the formula given above to obtain an approximate value of I where
    $$I = \int_0^{1/2} \frac{1}{\sqrt{(1-x^2)}} dx,$$
    giving your result to three decimal places. Show, by direct integration, that $I = \pi/6$. (L)

9. Use Simpson's rule with five ordinates (four strips) to find an approximate value of $\int_0^4 xe^{-x} dx$. Give your answer correct to two decimal places. (L)

10. Tabulate to three decimal places the values of the function $f(x) = \sqrt{(1+x^2)}$ for values of x from 0 to 0·8 at intervals of 0·1. Use these values to estimate $\int_0^{0·8} f(x) dx$ by Simpson's method (i) using all the ordinates, (ii) using only ordinates at intervals of 0·2. Draw any conclusions you can about the accuracy of the results. (L)

11 The velocity at time t seconds of a particle moving along the x-axis is given by $\sqrt{(8 - t^3)}\,\mathrm{m\,s^{-1}}$ for $\frac{1}{2} \leqslant t \leqslant 2$. Use Simpson's rule with six strips to estimate, correct to three significant figures, the average velocity of the particle during the interval $\frac{1}{2} \leqslant t \leqslant 2$. (L)

12 The coordinates of a point on a curve are given by the following table:

x	0	0·5	1·0	1·5	2·0	2·5	3·0
y	1	1·03	1·24	1·72	2·46	3·44	4·57.

Use Simpson's rule to show that the area of the region bounded by this curve, the axis Ox and the lines $x = 0$ and $x = 3$ is approximately 6·29 units^2. Calculate approximate coordinates for the centroid of this area. (L)

10.4 Approximate integration by series expansion of the integrand

An approximate value of a definite integral can sometimes be obtained by expanding the integrand in an infinite series and then integrating the series term by term. The success of this method depends upon the limits of the integral being such that they ensure rapid convergence of the integrated series within the required degree of accuracy.

Example 3 Evaluate, correct to four decimal places, $\displaystyle\int_0^1 \frac{\sin x}{x}\,\mathrm{d}x$.

Now $\quad \sin x = x - \dfrac{x^3}{3!} + \dfrac{x^5}{5!} - \dfrac{x^7}{7!} + \dfrac{x^9}{9!} - \cdots$

$\Rightarrow \dfrac{\sin x}{x} = 1 - \dfrac{x^2}{3!} + \dfrac{x^4}{5!} - \dfrac{x^6}{7!} + \dfrac{x^8}{9!} - \cdots$

$\Rightarrow \displaystyle\int_0^1 \frac{\sin x}{x}\,\mathrm{d}x = \int_0^1 \left[1 - \frac{x^2}{3!} + \frac{x^4}{5!} - \frac{x^6}{7!} + \frac{x^8}{9!} - \cdots \right]\mathrm{d}x$

$\qquad = \left[x - \dfrac{x^3}{3.3!} + \dfrac{x^5}{5.5!} - \dfrac{x^7}{7.7!} + \dfrac{x^9}{9.9!} - \cdots \right]_0^1$

$\qquad = 1 - \dfrac{1}{3.3!} + \dfrac{1}{5.5!} - \dfrac{1}{7.7!} + \dfrac{1}{9.9!} - \cdots$

$\qquad \approx 1 - 0·055\,556 + 0·001\,667 - 0·000\,028 + 0·000\,000\,3 - \cdots$

$\qquad \approx 0·946\,0833$

$\qquad = 0·9461$ (to 4 dec. places).

Example 4 Show by means of the binomial theorem that

$$\frac{1}{\sqrt{(1 - x^2)}} = 1 + \frac{x^2}{2} + \frac{3x^4}{8} + \frac{5x^6}{16} + \cdots$$

Hence, by considering $\int_0^{1/2} \dfrac{1}{\sqrt{(1-x^2)}}\,dx$ and working to five places of decimals, find an approximate value for π.

Using the binomial theorem, $\dfrac{1}{\sqrt{(1-x^2)}} = 1 + \dfrac{x^2}{2} + \dfrac{3x^4}{8} + \dfrac{5x^6}{16} + \ldots$

$$\Rightarrow \int_0^{1/2} \frac{1}{\sqrt{(1-x^2)}}\,dx = \int_0^{1/2} \left(1 + \frac{x^2}{2} + \frac{3x^4}{8} + \frac{5x^6}{16} + \ldots \right)dx$$

$$= \left[x + \frac{x^3}{6} + \frac{3x^5}{40} + \frac{5x^7}{112} + \ldots \right]_0^{1/2}$$

$$= \frac{1}{2} + \frac{1}{48} + \frac{3}{1280} + \frac{5}{14336} + \ldots$$

$$\approx 0.5 + 0.020\,83 + 0.002\,34 + 0.000\,35 + \ldots$$

$$\approx 0.523\,52.$$

But $\int_0^{1/2} \dfrac{1}{\sqrt{(1-x^2)}}\,dx = \left[\sin^{-1} x \right]_0^{1/2} = \sin^{-1} \tfrac{1}{2} - \sin^{-1} 0 = \pi/6,$

$$\Rightarrow \pi/6 \approx 0.523\,52 \Rightarrow \pi \approx 6 \times 0.523\,52 = 3.141\,12$$

$$\Rightarrow \pi = 3.141 \text{ (to 3 decimal places).}$$

Exercise 10.4

1 Find the value of $\int_0^1 \dfrac{1}{\sqrt[3]{x}} \ln\left(1 + \dfrac{x}{5}\right) dx$ correct to three places of decimals by expanding the integrand as a series in ascending powers of x.

2 By using suitable series expansions evaluate each of the following integrals correct to three decimal places:

(i) $\int_0^1 \sqrt{(4 + x^2)}\,dx$, (ii) $\int_0^{1/4} e^{\sqrt{x}}\,dx$, (iii) $\int_0^{1/10} \ln(1 + \sin x)\,dx$.

3 By means of the substitution $\sin x = t^2$, show that

$$I = \int_0^\alpha \sqrt{(\sin x)}\,dx = \int_0^{1/2} \frac{2t^2}{\sqrt{(1 - t^4)}}\,dt,$$

where $\sin \alpha = \tfrac{1}{4}$.

Estimate I to four places of decimals by expanding the integrand in ascending powers of t, retaining the first three non-zero terms and integrating term by term. All steps in the calculation are to be shown. (L)

4 Using Simpson's rule with four strips obtain an approximate value, to three decimal places, for $\int_0^{0.4} e^{-x^2}\,dx$. Check your result by expanding e^{-x^2} in ascending powers of x as far as the term in x^6 and integrating term by term. (L)

11 Differential equations

11.1 Introduction

Any equation involving the derivatives of one variable with respect to another variable is called a differential equation. Examples of a differential equation are (i) $\dfrac{dy}{dx} = 2x$, (ii) $\dfrac{dy}{dx} + 2xy = e^x$. Equations (i) and (ii) are said to be *first* order differential equations since the highest derivative each one contains is the *first*.

All simple integrals can be looked upon as the solution of a first order differential equation, for, if $\dfrac{dy}{dx} = f(x)$, then $y = \displaystyle\int f(x)\,dx$. Hence, the solution of the differential equation

$$\frac{dy}{dx} = 3\cos x + e^{2x}$$

is

$$y = \int (3\cos x + e^{2x})\,dx \Rightarrow y = 3\sin x + \tfrac{1}{2}e^{2x} + C,$$

where C is an arbitrary constant. This solution is said to be the general solution. If it is further known that when x has a particular value, say x_0, y has the value y_0, then C will automatically have a fixed value. For instance, if in the above equation it is also known that when $x = 0$, $y = 1$, then substituting these values in the general solution gives

$$1 = 3.0 + \tfrac{1}{2}.1 + C \Rightarrow C = \tfrac{1}{2}.$$

Hence $y = 3\sin x + \tfrac{1}{2}e^{2x} + \tfrac{1}{2}$. Such a solution is said to be a *particular solution*.

11.2 First order variables separable

Differential equations of the form $\dfrac{dy}{dx} = f(x)g(y)$ can be rearranged in the form $\dfrac{1}{g(y)}\dfrac{dy}{dx} = f(x)$. Integrating with respect to x gives

$$\int f(y)\frac{dy}{dx}dx = \int g(x)\,dx + C \Rightarrow \int f(y)\,dy = \int g(x)\,dx + C.$$

Such types in which all y's can be collected with the dy and all x's with the dx are called *separable differential equations*.

Example 1   Find the general solution of the differential equation

$$x\frac{dy}{dx} + y = x^2 y.$$

The given equation can be rearranged as

$$x\frac{dy}{dx} = y(x^2 - 1) \Rightarrow \int\frac{dy}{y} = \int\frac{(x^2 - 1)}{x}dx$$

$$= \int\left(x - \frac{1}{x}\right)dx.$$

$$\Rightarrow \ln y = \tfrac{1}{2}x^2 - \ln x + C.$$

Example 2   Solve the differential equation $y\sqrt{(1 - x^2)}\dfrac{dy}{dx} = \dfrac{1}{y} - y^2$, given

that $y = \dfrac{1}{2}$ when $x = 0$.

$$y\frac{dy}{dx} = \frac{1}{\sqrt{(1 - x^2)}}\cdot\frac{1 - y^3}{y} \Rightarrow \int\frac{y^2}{1 - y^3}dy = \int\frac{1}{\sqrt{(1 - x^2)}}dx$$

$$\Rightarrow -\frac{1}{3}\ln(1 - y^3) = \sin^{-1}x + C.$$

When $x = 0$, $y = \tfrac{1}{2}$, $\Rightarrow C = -\tfrac{1}{3}\ln(1 - \tfrac{1}{8}) - \sin^{-1} 0 = -\tfrac{1}{3}\ln\tfrac{7}{8}$

$$\Rightarrow \ln(1 - y^3) = \ln(7/8) - 3\sin^{-1}x.$$

Exercise 11.2

1   Find the general solution of the differential equations

(i) $\dfrac{dy}{dx} = \dfrac{2 - y}{1 + 2x}$,   (ii) $xy\dfrac{dy}{dx} = 1 - y^2$,   (iii) $\cos^2 t\dfrac{dy}{dt} = e^{2y}$,

(iv) $y - x\dfrac{dy}{dx} = 2\left(y^2 + \dfrac{dy}{dx}\right)$,   (v) $\dfrac{dx}{dt} = e^{t + 2x}$.

2   Solve the differential equation

$$(1 + t^2)\frac{d\theta}{dt} - \theta(\theta + 1)t = 0,$$

given that $\theta = 1$ when $t = 0$.

3   Solve the differential equation

$$\cos y\frac{dy}{dx} = \cot x(1 + \sin y),$$

given that $y = \pi/2$ when $x = \pi/2$.   (AEB)

4 Solve the differential equation $\cos x \dfrac{dy}{dx} - \sqrt{y} \sin x = y \sin x$, given that $y = 4$ when $x = 0$.

(AEB)

5 Find y as a function of x when

$$(1 + x^2)\frac{dy}{dx} = x(1 - y^2),$$

given that $y = 0$ when $x = 1$.

(L)

6 Find the equation of the curve which passes through the origin and satisfies the differential equation

$$\frac{dy}{dx} = \frac{x}{1 + y^2}.$$

Show that the origin is a stationary point and find whether it is a maximum, minimum, or point of inflexion.

(L)

7 Solve the differential equation

$$2\frac{dy}{dx} + xy^2 = y^2,$$

given that $y = -1$ when $x = 1$. Show that the solution may be expressed in the form

$$y = \frac{4}{(x - 3)(x + 1)}.$$

Sketch the graph of y against x for $-1 < x < 3$, showing the asymptotes and giving the coordinates of any turning point.

(L)

8 Find the equation of the curve which passes through the point $(0, 1)$ and which satisfies the differential equation

$$\frac{dy}{dx} = y\sqrt{(1 + x)}.$$

9 A time T is allowed for a journey of length a. The speed decreases with the time from the initial speed in such a way that, if after time t and at distance x from the start the speed becomes constant with the value it then had, the journey would be completed in half the remaining time available. Show that the speed v at time t is given by

$$v = \frac{2(a - x)}{T - t}.$$

Solve this differential equation for x in terms of t and show that the acceleration is constant.

(L)

10 The variable x satisfies the differential equation

$$\frac{dx}{dt} = a(1 - x)(2 - x) - bx^2,$$

where a and b are constants. Show that, if the limiting value of x as t increases indefinitely is $\frac{1}{2}$, then $b = 3a$. Given that $a = \frac{1}{5}$ and that $x = 0$ when $t = 0$, find x in terms of t. Find also the value of t when $x = 0.4$. Find, with the same values of a and b, an expression for x in terms of t when $x = 1$ at $t = 0$.

(L)

Answers

In all cases, the constants of integration have been omitted from answers to indefinite integrals.

Exercise 1.3
1   $\frac{1}{3}(7 + x)^3$
2   $-\frac{1}{3}(3 - 2x)^{3/2}$
3   $-\frac{3}{4}(1 - t)^{4/3}$
4   $\frac{1}{4}(1 + \frac{x}{2})^8$
5   $\frac{2}{19}(3x - 2)^{19/6}$
6   $-\frac{2}{3}(3 - 2u)^{3/4}$
7   $-\ln(1 - x)$
8   $\frac{1}{2}\ln(2x + 3)$
9   $-\frac{1}{2}\cos 2x$
10   $\sin t$
11   $\frac{1}{3}\cos 3x$
12   $\frac{1}{4}\sin 4x$
13   $2\sin \frac{1}{2}t$
14   $\frac{1}{4}\tan 4x$
15   $-\frac{4}{3}\tan 3(\pi - \frac{u}{4})$
16   $\frac{1}{2}e^{2x}$
17   $-\frac{1}{2}e^{1-2x}$
18   $-\frac{1}{3}e^{-3u}$
19   $\frac{1}{2}(e^{2x} - e^{-2x}) + 2x$
20   $4x + \frac{4}{3}e^{3x} + \frac{1}{6}e^{6x}$

Exercise 1.4
1   $\sin^{-1}(x/2)$
2   $\sin^{-1}[(x + 1)/2]$
3   $\frac{1}{3}\sin^{-1}(3x/5)$
4   $\frac{1}{10}\tan^{-1}(x/10)$
5   $\frac{1}{10}\tan^{-1}(10x)$
6   $\dfrac{1}{\sqrt{15}}\tan^{-1}(x\sqrt{3}/\sqrt{5})$
7   $\sin^{-1}[(x - 3)/2]$
8   $\frac{1}{8}\tan^{-1}[(2x + 1)/4]$

Exercise 2.1
1   $\frac{1}{2}\ln(1 + \sin 2x)$
2   $\frac{1}{6}\sin^6 x$
3   $\frac{1}{2}(\sin^{-1}x)^2$
4   $\frac{1}{4}\ln(1 + 2x^2)$
5   $-\frac{1}{4}\sqrt{(9 - 4x^2)}$
6   $\frac{1}{3}(\ln x)^3$
7   $\ln(\ln x)$
8   $\frac{1}{3}\ln(e^{3x} - 1)$

9   $1/(1 - \tan x)$
10   $-2\cos\sqrt{x}$
11   $\frac{2}{3}(x - 4)^{3/2} + 8(x - 4)^{1/2}$
12   $\dfrac{1}{8\cos^2 4x}$
13   $\frac{1}{4}\tan^{-1}x^4$
14   $\frac{1}{12}(1 + 4x^2)^{3/2}$
15   $2\sqrt{(\sin x)}$
16   $-(1 + e^x)^{-2}/2$
17   $-e^{\cos x}$
18   $-e^{\cos^2 x}$
19   $2(\sin x + 1)^{1/2}$

Exercise 2.2
1   $-\sqrt{(1 - x^2)}$
2   $\dfrac{x}{4\sqrt{(4 + x^2)}}$
3   $\frac{2}{3}(x - 1)^{3/2} + \frac{2}{5}(x - 1)^{5/2}$
4   $2/(1 - x)^{1/2}$

Exercise 2.3
1   $\frac{1}{2}(\ln x)^2$
2   $\frac{1}{6}\tan^2 3x$
3   $\frac{3}{4}(x^2 + 1)^{2/3}$
4   $\frac{1}{5}\ln(1 + 5x^2)$
5   $\ln(1 + e^x)$
6   $-\frac{1}{10}(1 - e^{2x})^5$
7   $\frac{1}{4}\sin^4 x$
8   $-\frac{1}{2}\ln\cos 2x$
9   $\dfrac{1}{3}\sin^{-1}\left(\dfrac{x^3}{3}\right)$
10   $\frac{1}{2}(\sin^{-1}x)^2$
11   $-\frac{1}{2}\ln(1 - \tan 2x)$
12   $\dfrac{-1}{(1 + \ln x)}$

Exercise 3.2
1   $\frac{1}{4}\sin^2 2x$
2   $\frac{1}{4}\sin 2t - \frac{1}{20}\sin 10t$
3   $\frac{1}{4}\sin 2x + \frac{1}{24}\sin 12x$
4   $\frac{1}{8}\cos 4x - \frac{1}{12}\cos 6x$

5 $-x + \frac{1}{2}\tan 2x$
6 $\frac{1}{2}\ln \tan (\frac{\pi}{4} + x)$
7 $\tan t$
8 $\frac{x}{2}\sin 2 + \frac{1}{4}\cos 2x$
9 $\frac{1}{3}\ln \sin 3t$

Exercise 3.3
1 $-\cos x + \frac{1}{3}\cos^3 x$
2 $\dfrac{3x}{8} - \dfrac{1}{4}\sin 2x + \dfrac{1}{32}\sin 4x$
3 $-\frac{1}{3}\cos^3 x + \frac{1}{5}\cos^5 x$
4 $\tan x + \frac{2}{3}\tan^3 x + \frac{1}{5}\tan^5 x$
5 $\frac{1}{5}\tan^5 x$
6 $\sec x - \frac{2}{3}\sec^3 x + \frac{1}{5}\sec^5 x$
7 $\frac{1}{3}\tan^3 x - \tan x + x$
8 $-\cot x - \frac{1}{3}\cot^3 x$
9 $-\cot (\frac{x}{2})$
10 $-2/(1 + \tan \frac{x}{2})$
11 $\frac{1}{4}\tan^{-1}\left(\dfrac{\tan x}{4}\right)$
12 $\dfrac{2}{\sqrt{3}}\tan^{-1}\left(\dfrac{1 + 2\tan x}{\sqrt{3}}\right)$

Exercise 4.2
1 $2x - 8\ln (x + 4)$
2 $-\dfrac{x^2}{2} - 3x - 9\ln (3 - x)$
3 $-\dfrac{2x}{3} - \dfrac{5}{9}\ln (1 - 3x)$
4 $\dfrac{x^2}{2} - \dfrac{2x}{3} + \dfrac{22}{9}\ln (3x + 2)$

Exercise 4.3
1 $\ln\left(\dfrac{x - 3}{x + 2}\right)$
2 $\frac{7}{9}\ln (x + 5) + \frac{11}{9}\ln (x - 4)$
3 $-\dfrac{x}{4} + \dfrac{5}{16}\ln\left(\dfrac{1 + 2x}{1 - 2x}\right)$
4 $\ln (x - 1) - \dfrac{3}{x - 1}$
5 $\frac{2}{3}\ln x - \frac{1}{15}\ln (5x + 3)$
6 $-\dfrac{x^2}{2} - \dfrac{9}{2}\ln (9 - x^2)$
7 $\frac{5}{24}\ln (3x + 1) + \frac{9}{8}\ln (x + 3)$

Exercise 4.4
1 $\tan^{-1}(x + 2)$
2 $\dfrac{2}{\sqrt{7}}\tan^{-1}\left(\dfrac{4x + 3}{\sqrt{7}}\right)$
3 $\dfrac{\sqrt{3}}{4}\ln\left(\dfrac{2 + x\sqrt{3}}{2 - x\sqrt{3}}\right)$

4 $\dfrac{1}{\sqrt{5}}\ln\left(\dfrac{\sqrt{5} + x + 1}{\sqrt{5} - x - 1}\right)$

Exercise 4.5
1 $\frac{3}{2}\ln (x^2 - 5) + \dfrac{1}{\sqrt{5}}\ln\left(\dfrac{x - \sqrt{5}}{x + \sqrt{5}}\right)$
2 $\ln (2x^2 + 2x + 1) - \tan^{-1}(2x + 1)$
3 $x + \ln (x^2 - 2x + 5)$
$- 2\tan^{-1}\left(\dfrac{x - 1}{2}\right)$
4 $-\dfrac{1}{2}\ln (1 + 4x - 2x^2)$
$- \dfrac{7}{2\sqrt{6}}\ln\left(\dfrac{\sqrt{2}(x - 1) - \sqrt{3}}{\sqrt{2}(x - 1) + \sqrt{3}}\right)$

Exercise 4.5
1 $\ln\left(\dfrac{x^2 - 1}{x^2}\right)$
2 $\frac{1}{5}\ln (x + 2) - \frac{1}{10}\ln (x^2 + 1) + \frac{2}{5}\tan^{-1}x$
3 $\tan^{-1}(x + 1) - \ln (x + 1) + \frac{1}{2}\ln (x^2 + 2x + 2)$
4 $\dfrac{1}{12}\ln (x - 2) - \dfrac{1}{24}\ln (x^2 + 2x + 4)$
$- \dfrac{1}{4\sqrt{3}}\tan^{-1}\left(\dfrac{x + 1}{\sqrt{3}}\right)$

Exercise 5.2
1 $\sin^{-1}\left(\dfrac{x}{2}\right) - \sqrt{(4 - x^2)}$
2 $7\sin^{-1}\left(\dfrac{x - 2}{3}\right) - 2\sqrt{(5 + 4x - x^2)}$
3 $3\sin^{-1}\left(\dfrac{x - 1}{3}\right) + \sqrt{(8 + 2x - x^2)}$

Exercise 6.1
1 $\frac{1}{9}\sin 3x - \frac{x}{3}\cos 3x$
2 $-(1 + \ln x)/x$
3 $\left(\dfrac{x^2}{4} - \dfrac{x}{8} + \dfrac{1}{32}\right)e^{4x}$
4 $(x + 2)\ln (2 + x) - x$
5 $x\sin^{-1} x + \sqrt{(1 - x^2)}$
6 $e^x(\cos 2x + 2\sin 2x)/5$
7 $\frac{1}{2}x\sqrt{(1 - x^2)} + \frac{1}{2}\sin^{-1} x$
8 $x(\ln x)^2 - 2x\ln x + 2x$
9 $\dfrac{2x}{3}(1 + x)^{3/2} - \dfrac{4}{15}(1 + x)^{5/2}$
10 $\dfrac{x^2}{4} - \dfrac{x}{4}\sin 2x - \dfrac{1}{8}\cos 2x$

Exercise 7.2

1 $\dfrac{128\sqrt{2}}{105}$

2 $\frac{16}{3} - 2\ln 3$

3 $\frac{8}{3} - 2^{3/4} - \frac{1}{3}.2^{1/4}$

4 $\frac{4}{15}(\sqrt{2} + 1)$

5 $\frac{4}{5}$

6 $\frac{3}{2} - \frac{1}{2}e^{-1}$

7 $\frac{3}{10}$

8 $\frac{2}{15}$

9 $\dfrac{\pi}{24} + \dfrac{\sqrt{3}}{64}$

10 $\frac{\pi}{4}$

11 (i) $\frac{16}{15}$, (ii) $\frac{\pi}{8} + \frac{1}{4}$, (ii) $7\frac{11}{35}$

12 $\frac{1}{3}\tan^{-1} 3 - \frac{1}{18} + \frac{1}{162}\ln 10$

Exercise 8.1

1 $y = Ax^{3/2} + B$

2 $4\frac{2}{3}$

3 $2y = 3x^2 - 1$

4 $y = Ke^{x/c}$

5 $3\text{ s}, 31\frac{1}{2}\text{ m}$

6 $y = \dfrac{7x^2}{2} + \dfrac{x^3}{3} + \dfrac{2}{3}\sin 3x - x + 2$

7 $y = \dfrac{1}{2}\left(\dfrac{1}{2 - x}\right) + 2x - \dfrac{1}{2}$

Exercise 8.4

1 1

2 $\dfrac{8\sqrt{2}}{3}$

3 $57\frac{1}{6}$

4 $y = 3x^2 - x^3, 6\frac{3}{4}$

5 $\frac{1}{4}(2 + \sqrt{3})$

6 $\dfrac{2}{\sqrt{5}}\tan^{-1}\dfrac{2}{\sqrt{5}} - \dfrac{1}{2}\ln\left(\dfrac{9}{5}\right)$

8 $24\sqrt{3}/5$

9 $y = 3x - 2, 6\frac{3}{4}$

Exercise 8.6

1 $(0, 0)$ and $(4a, 4a)$

2 $\pi r^3/6$

3 22.2 cm^3

4 $16a^5/15, \pi a^6/3$

5 $\pi(4h + 4e^h + \frac{1}{2}e^{2h} - 4\frac{1}{2}), 0.379\text{ s}^{-1}$

6 (i) $\pi^2/2$, (ii) $2\pi^2$

7 $\pi a^2(4\ln 2 - 3/2)$

Exercise 8.10

1 $k = \pm[(16 - 3\pi)/(3\pi)]^{1/2}$

2 $2/(3\pi) : 1/\sqrt{2}$

3 (i) $\frac{1}{2}\ln(\frac{8}{3})$, (ii) 1.45

4 $3\pi, \frac{5}{6}$

5 $y = a - \frac{1}{2}x, y = a + 2x, 6a/5$

6 $0.0452, 0.257$

7 $(\pi, \frac{5}{6})$

8 $3.53, 4.97$

Exercise 9.1

1 (a) 0, (b) 2, (c) $\frac{4}{35}$, (d) 0

2 $\frac{2}{3}$

3 $\frac{4}{15}$

4 π

5 (i) $(-\text{ve})$, (ii) 0, (iii) $(-\text{ve})$

6 (a) $\dfrac{\pi^2}{2} - 2$, (b) $\dfrac{5\pi^2}{8} - 3$

7 (a) $\dfrac{1}{3}$, (b) $(e^2 + 1)/4$,

   (c) $\dfrac{\pi}{3} + \sqrt{3} - 2$

Exercise 10.3

1 $0.5246, 3.148$

2 1.1484

3 1.769

4 $5.354, 2.677$

5 (i) 16.67, (ii) $(3.015, 1.576)$

6 $11.188, 32.831$

7 0.019

8 0.524

9 0.89

10 (i) 0.879, (ii) 0.879

11 2.21

12 $(1.93, 1.33)$

Exercise 10.4

1 0.113

2 (i) 2.080, (ii) 0.351, (iii) 0.005

3 0.0845

4 0.380

Exercise 11.2

1 (i) $(1 + 2x)(2 - y)^2 = C$,
   (ii) $x^2(1 - y^2) = C$,
   (iii) $e^{-2y} + 2\tan t = C$,
   (iv) $y/(1 - 2y) = C(x + 2)$
   (v) $e^{-2x} = C - 2e^t$

2 $2\theta/(1 + \theta) = \sqrt{(1 + t^2)}$

3 $1 + \sin y = 2\sin x$

4 $(1 + \sqrt{y})^2 = 9\sec x$

5 $2(1 + y)/(1 - y) = 1 + x^2$

6 $6y + 2y^3 = 3x^2$, minimum

8 $3\ln y = 2(1 + x)^{3/2} - 2$

9 $T^2(a - x) = a(T - t)^2$

10 $x = \dfrac{2(e^t - 1)}{4e^t + 1}, t = \ln 6$,

    $x = \dfrac{2 + 3e^t}{6e^t - 1}$

Index

acceleration   31, 32
algebraic fractions
  irrational   23–4
  rational   18–22
approximate integration   63–6
area under a curve   33–42, 63–6
  first moment of   51–3
  negative   37–41

binomial theorem   69

centroid   54–7
constant of integration   1, 31, 70

definite integrals   28–30, 58–62
difference of functions   2
differential equations   70–71

gradient   31

limits   28, 58
logarithmic functions   1, 27

mean values   49–50
moment of area   51–3

notation, alternative   10–11

parametric form of curve   41–2
partial fractions   19, 50
parts, integration by   25–7

root mean square value   51

series expansion   68–9
shell method   47–9
Simpson's rule   65–6
standard integrals   1, 4, 5, 13
substitution   7–12, 28–30
sum of functions   2

tan half-angle   16
trapezoidal rule   63–4
trigonometric functions   4, 13–17
  inverse   5, 27

velocity   31, 32
volume of revolution   43–7
  by shell method   47–9